AF769135

Biocomposites

Edited by Brajesh Kumar

Published in London, United Kingdom

IntechOpen

Supporting open minds since 2005

Biocomposites
http://dx.doi.org/10.5772/intechopen.92504
Edited by Brajesh Kumar

Contributors
Brajesh Kumar, Harekrishna Mahato, Ashwani Sharma, Nagashree Nagaraj Rao, Kalmanje Mugdha Bhat, Jyothsana Rajagopalan, Rajeshwari Mallikarjunaiah, Ahmed Koubaa, Mourad Saddem, Bernard Riedl, Sushant Kumar, Swarnima Pandey, NV Satheesh Madhav, Basel Younes, Pradeep Sharma, Ahmed Sharif, MD Rajbanul Akhond

© The Editor(s) and the Author(s) 2022
The rights of the editor(s) and the author(s) have been asserted in accordance with the Copyright, Designs and Patents Act 1988. All rights to the book as a whole are reserved by INTECHOPEN LIMITED. The book as a whole (compilation) cannot be reproduced, distributed or used for commercial or non-commercial purposes without INTECHOPEN LIMITED's written permission. Enquiries concerning the use of the book should be directed to INTECHOPEN LIMITED rights and permissions department (permissions@intechopen.com).
Violations are liable to prosecution under the governing Copyright Law.

CC BY

Individual chapters of this publication are distributed under the terms of the Creative Commons Attribution 3.0 Unported License which permits commercial use, distribution and reproduction of the individual chapters, provided the original author(s) and source publication are appropriately acknowledged. If so indicated, certain images may not be included under the Creative Commons license. In such cases users will need to obtain permission from the license holder to reproduce the material. More details and guidelines concerning content reuse and adaptation can be found at http://www.intechopen.com/copyright-policy.html.

Notice
Statements and opinions expressed in the chapters are these of the individual contributors and not necessarily those of the editors or publisher. No responsibility is accepted for the accuracy of information contained in the published chapters. The publisher assumes no responsibility for any damage or injury to persons or property arising out of the use of any materials, instructions, methods or ideas contained in the book.

First published in London, United Kingdom, 2022 by IntechOpen
IntechOpen is the global imprint of INTECHOPEN LIMITED, registered in England and Wales, registration number: 11086078, 5 Princes Gate Court, London, SW7 2QJ, United Kingdom
Printed in Croatia

British Library Cataloguing-in-Publication Data
A catalogue record for this book is available from the British Library

Additional hard and PDF copies can be obtained from orders@intechopen.com

Biocomposites
Edited by Brajesh Kumar
p. cm.
Print ISBN 978-1-83969-080-8
Online ISBN 978-1-83969-081-5
eBook (PDF) ISBN 978-1-83969-082-2

For EU product safety concerns:
IN TECH d.o.o., Prolaz Marije Krucifikse Kozulić 3, 51000 Rijeka, Croatia, info@intechopen.com or visit our website at intechopen.com.

We are IntechOpen, the world's leading publisher of Open Access books

Built by scientists, for scientists

5,700+
Open access books available

140,000+
International authors and editors

175M+
Downloads

156
Countries delivered to

Our authors are among the
Top 1%
most cited scientists

12.2%
Contributors from top 500 universities

WEB OF SCIENCE™

Selection of our books indexed in the Book Citation Index (BKCI)
in Web of Science Core Collection™

Interested in publishing with us?
Contact book.department@intechopen.com

Numbers displayed above are based on latest data collected.
For more information visit www.intechopen.com

Meet the editor

Dr. Brajesh Kumar is currently an assistant professor and head in the Department of Chemistry, TATA College, Kolhan University, Chaibasa, India. He received a Ph.D. in Chemistry from the University of Delhi, India, in 2009. His research interests include the development of sustainable and eco-friendly techniques for nanoparticle synthesis and their applications for environmental remediation, nanofibers, nanocomposites, nanomedicine, sensors, active films of organic solar cells, natural product extraction, purification and analysis, natural polymers, peptide chemistry, microwave- and ultrasound-assisted organic synthesis, and organic synthesis. Dr. Kumar has several national and international fellowships to his credit and has worked as a faculty member in various universities throughout India, Ecuador, and South Korea. He has also published numerous research articles and is an active reviewer of more than seventy journals. He is also included in the top 2% of scientists (2021) ranked by Stanford University, USA, and Elsevier.

Contents

Preface

Biocomposites are natural fibre-reinforced biopolymers or synthetic fibres that remain bonded together by physical or chemical interactions but retain their individual physical or chemical identities. Their melting point and viscosity are influenced by the type of polymer matrix, while their mechanical properties are governed to a large extent by the type and amount of filler. Their applications are based on their physical, mechanical, and processing properties. Nowadays, many investigators have developed these materials via eco-friendly methods using lignocellulose, keratin, silk, rice straw, paper sludge, coconut coir, polylactic-co-glycolic acid, alginate, chitosan, graphene, gelatin, dextran, starch, gum, and so on as alternatives to conventional materials due to their sustainability, nontoxic nature, biodegradability, low cost, and more. In this book, researchers from all over the world highlight the importance of biocomposites and natural products.

Green composites are promising because they are renewable, biodegradable, and durable for non-renewable composites. In Chapter 1, Bhat et al. discuss green biocomposites and their importance. In Chapter 2, Sharma discusses the opportunities of extractive, cellulosic, and lignocellulosic fibres from non-wood forest products in biocomposites. In Chapter 3, Akhond and Sharif demonstrate the function of biocomposites in various engineering and biomedical materials. In Chapter 4, Sushant et al. examine the ability of biopolymers isolated from natural sources as novel bioexcipients in the design of novel drug delivery. In Chapter 5, Saddem and Koubaa present the properties of high-density polyethylene-polypropylene wood composites. In Chapter 6, Basel reviews the characterization, modelling, and production processes of biopolymers in the textile industry, which is an exciting and innovative area of research. In the final chapter, Harekrishna and Brajesh discuss the pharmacological applications and nutritional properties of *Aegle marmelos*, a rare species of tree also known as bael.

I am grateful to my wife, Kumari Smita, for her helpful comments on several chapters and excellent support in conceptualizing this book.

Dr. Brajesh Kumar
Department of Chemistry,
TATA College,
Kolhan University,
Chaibasa, Jharkhand, India

Chapter 1

Eco-Friendly and Biodegradable Green Composites

Kalmanje Mugdha Bhat, Jyothsana Rajagopalan, Rajeshwari Mallikarjunaiah, Nagashree Nagaraj Rao and Ashwani Sharma

Abstract

Natural fibers, are environmental friendly, biodegradable, abundantl, renewable and cheap with low density. Plant fibers are light compared to glass, carbon and aramid fibers. The biodegradability of plant fibers contribute to a healthy ecosystem while their low cost and high performance fulfills the economic interest. The effect of fiber content on the properties of natural fiber reinforced composites is particularly significance. Important factor that significantly influences the properties and interfacial characteristics of the composites is the processing parameters used. Biocomposites offers a significant market in automotive and decking market but application in other sectors has been limited. Green composites are promising because they are renewable, biodegradable and sustainable for non-renewable composites.

Keywords: biocomposite, biodegradable, eco-friendly, sustainable, fibers

1. Introduction

Globalization and sustainability has made life not only feasible but challenging too. Materials which are obtained from resources that are renewable tend to be suitable for sustainable development. These materials have a global value since they can act as a counter to the various environmental issues such as waste management problems, increase in global warming, the constant rise in oil prices and the deteriorating fossil resources. Different varieties of renewable materials have been used for many years across the food, furniture, and textile industry such as vegetable oils, starch and cellulosic based polymers, cotton, natural fibers, silk, and wool [1]. On the other hand, it is only recently that these materials have gained interest as a potential alternative to synthetic based polymers for different kinds of industrial applications like automotive, films, construction, paper coating, packaging and biomedical applications. The synthetic polymers pose many drawbacks towards the environment in ways such as the amount of vapors and toxic gases released after incineration and improper disposal, there has been more research work being focused on new green biopolymeric materials and their effective utilization in green composite applications.

Over the years, bioproducts have gained commercial importance. Chemical processes such as production of 'green' ethylene through dehydration of ethanol and further production of 'green' polyethylene, polyvinyl chloride and some other plastics have been reviewed. Certain technological developments have also been used to enhance certain material properties of polymers that are bio-based; an example of which is development of heat resistant polylactic acid, thereby allowing extensive applications. Bio-fibers with stable properties are being produced over time by optimizing plants. There have been numerous applications occurring lately such as packaging, biomedical products, textile, agriculture, construction where these biodegradable biopolymers and biocomposites are an appropriate sustainable replacement [1, 2].

Biospecific and biosimulation materials cover the whole field of biofunctional materials. Biofunctional materials are synthesized from the view point of functionality design. The functionality design is based on determination of the polymer structure that realizes the desired functionality and property of materials, and on exploration of the appropriate method of polymer synthesis, polymer reaction, and polymer modification that yields the designed polymer structure.

1.1 Concept of composites

A composite is a structural material which includes a combination of different entities that are insoluble in each other and are mixed together at a macroscopic level [1]. One of the constituents is known as the reinforcing phase and the other one into which the reinforcing phase is embedded is called the matrix (**Figure 1**). The reinforcing phase materials are found to be made of varying textures that can be in the form of flakes, fibers or particles [2]. On the other hand, the matrix phase materials are generally made of continuous phases [3, 4]. One of the most common examples of naturally found composites includes wood (cellulose fibers are reinforced into lignin matrix) and bones (reinforcements of bone-salt plates consisting of phosphate and calcium ions are added to the soft collagen matrix). Each constituent of the composite has different roles giving rise to a strong structural material. The matrix component within composite materials gives a defined shape, protects the reinforcements from environmental damage, transfers loads to reinforcing phase and improves the toughness of material [5]. The reinforcements in composites get strength from the matrix, stiffness and other mechanical properties; contain a high thermal expansion coefficient, high conductivity and good thermal transport [6].

1.2 Pros and cons of composites materials

1.2.1 Pros of composites

Composite materials have various advantages that can be used for different applications. Composite materials are very lightweight and have a low density

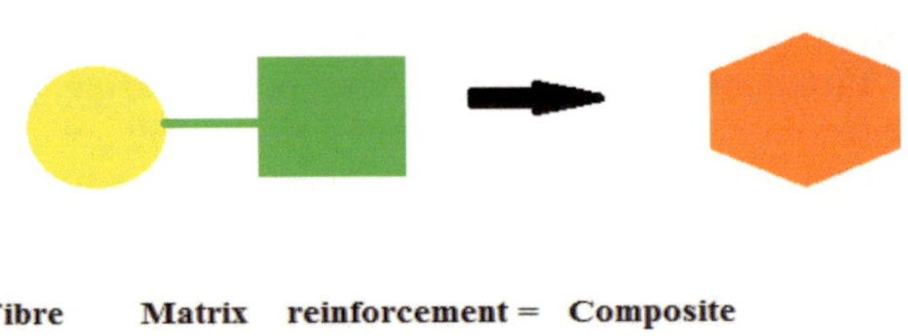

Figure 1.
Concept of biocomposite.

making it easier to mold into complex forms and modules like Modular construction. These materials have a high specific stiffness, creep resistance and strength with an improved friction and wear properties. Composites have low thermal expansion and electrical conductivity giving rise to good dumping properties and good fatigue resistance. These composite materials are easily bondable, have lower radar visibility and have the capability to store and release internal energy, leading to a lower overall system cost. Composite products are known to have excellent heat sink properties. The mechanical properties such as strength-weight ratio, stiffness-weight ratio and fatigue properties of the composite products are better than most of the common engineering metals like steel or aluminum. With improved corrosion resistance and a high resistance to impact damage, the composites have indefinite shelf life just like metals. Composites can be used to obtain a varied combination of properties which cannot be attained in other materials such as polymers, ceramics or metals alone. One of the major advantages of composites is the ability to incorporate sensors into the matrix which can be used to monitor and correct the material performance, giving rise to a range of applications towards Smart composites (**Table 1**).

1.2.2 Limitations of composites

Though composite materials possess a huge number of advantages, there are certain limitations to the same. The composite materials are known to have a low ductility and have certain temperature limits. Due to the presence of more than one component, the material can be subjected to solvent or moister attack. The matrix component of the composite is generally weak giving rise to low toughness, higher brittleness, higher susceptibility to damage and weak transverse properties. The

The properties of bio composites	**Characteristics**		
	Mechanical	**Physical**	**Chemical/ Biological**
Intrinsic properties (based on chemical composition)	Flow limit Tensile/ compression resistance Poisson ratio Electricity modulus	Density Form and geometry Color esthetics	Flow limit Tensile/ compression resistance Poisson ratio Electricity modulus
Behaviour	Breaking strength Fatigue resistance Crack resistance Rigidity Wear resistance Shock resistance	Coefficient of thermal expansion Electrical conductivity Refractive index	Biofunctionality Bioinert Bioactive Biostability Biodegradation
Surface properties	Bending modulus Hardness Shearing modulus Bending resistance Shearing resistance	Surface topology Texture Roughness Hardness Coefficient of friction	Adhesion
Processing	Reproducibility, can be sterilized, Packaging feature		

Table 1.
Biocomposite properties different categories.

composite materials are difficult to attach onto other materials, requiring the need of additional materials such as fasteners. Repairing composites can introduce many other problems, making the disposal and reuse of composites very difficult. Composites have a limited shelf-life discouraging its usage for long term applications. Composite materials have a very complicated manufacturing and fabrication process requiring sophisticated tools and high cost raw materials. Composite materials require refrigerated transport and storage. In most of the cases, hot curing is necessary which takes time and requires special tools; overall adding to the high expense of producing composite materials.

1.3 Difference between smart and composite material

Though smart materials and composite materials sound similar, they are fundamentally different. Smart materials have a minimum dual function whereas composite materials consist of dual or more components/phases (thereby being called hybrid materials).

Smart materials serve various functions within one product, some common instances of which include actuator/sensor ability in addition to having form, having the ability to bear structural weight without breaking. The classic example having huge industrial value is Nitinol, an alloy of Nickel-Titanium. This material can be mechanically deformed like bent and can be returned to the pre-deformed shape by heating the material. Another good example of a smart material will be Lead-Zirconate-Titanate (PZT) which is a ceramic that mechanically deforms when an electrical potential is generated. PZT alloy has found an actuator application because a reverse potential produces expansion of the material geometrically (**Figure 2**).

As mentioned previously, composites consist of many different materials in two or more phases, which allow us to specifically engineer the composites to produce desired properties like enhancing mechanical stress, conductivity, etc. The

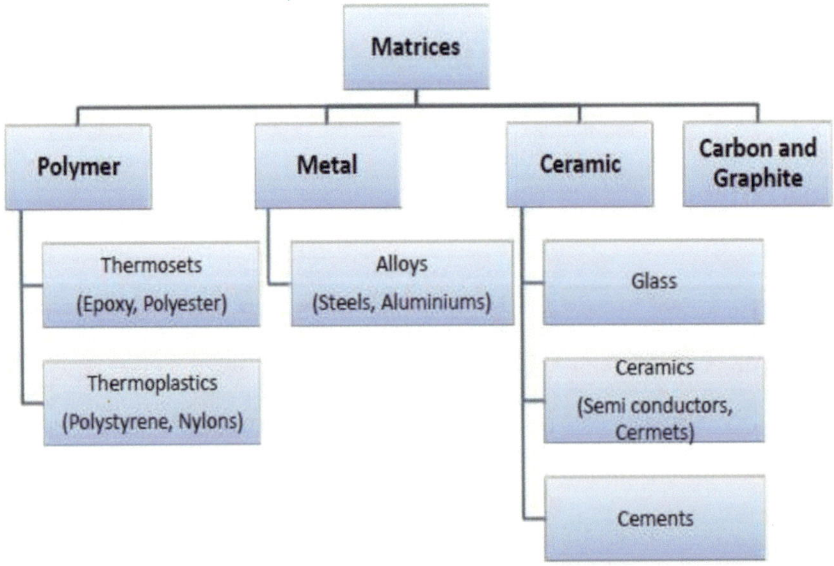

Figure 2.
Classification of composite materials with respect to matrices.

compound will be considered a composite material as long as the base materials retain their physical morphology and characteristics.

1.4 Applications of biocomposites

Electrical, Electronics	• Antennas • Cable tracks, Windmills • Insulation for electrical construction • Circuit breakers, Printed circuits • Armors, boxes • Television towers - Top
Buildings and Public Works	• Facade panels • Covers (domes, windows) • Chimneys • Profiles, Partitions • Swimming pool • Concrete molds
Road and Rail Transports	• Casings, cabins • Panels • Isothermal trucks • Body components • Power units, wagons, doors, seats • Bottles for gas • Wheels, grills • Ventilation housings • Chassis, Suspension • Highway tankers • Shafts and suspension springs • Trailers
Marine and Cable Transports	• Boats: Pleasure and racing canoes • Crafts, Patrol boats, Trawlers • Cabins: Telecabins, Telepherique • Anti-mine ships, Landing gears
Air and Space Transports	• Rocket boosters • Shields • Passenger aircrafts and Gliders • Reservoirs, Nozzles • Blades, Propellers • Shafts and Brake discs • Aircraft components: radomes, ailerons, stabilizers
Mechanical applications	• Weaving machine rods • Compressed gas bottles, Tubes for offshore platforms • Gears, Bearings, Casings • Pneumatics for radial frames • Jack body, Robot arms, Flywheels • Pipes, Components of drawing table
Sports and Recreation	• Skates, Bows and arrows • Rackets, Fishing poles • Javelins, Helmets • Skis, Poles used in jumping • Sails, Surf boards

2. Constituents of composites

Biocomposite materials are a hybrid of a reinforcement material made of natural fibers (plants or from cellulose derivatives) and a matrix material (resin). The resin is generally a polymer matrix made of either renewable or non-renewable resources. The natural fibers, also called biofibres, can be made from wood fibers (softwood and hardwood) or from non-wood fibers (hemp, wheat, flax, jute, sisal, kenaf, etc).

2.1 Classification of composites materials

The classification of composite materials can be done into different classes based on the phases:

- One of the most commonly used classifications is based on the material of the matrix. The major classes include Organic Matrix Composites (OMCs), Ceramic Matrix Composites (CMCs) and Metal Matrix Composites (MMCs).

- The class Organic Matrix Composite is a more general term which includes subclasses, including Carbon matrix composites or carbon–carbon composites and Polymer Matrix Composites (PMCs).

Another commonly used classification is based on the material of the reinforcement used. The classes are namely fiber reinforced composites, particulate composites, and structural composites.

- Fiber Reinforced Composites, as the name suggests, are made from fibers that are embedded into the matrix material. The Fiber Reinforced composites (FRP) can be further classified into 2 types - Short-fiber reinforced composites (those containing discontinuous fibers) and Long-fiber reinforced composites (those with continuous fibers). A fiber composite can be called a discontinuous one when the properties of the material vary according to the fiber length. On the other hand, the composite can be called continuous fiber if further increase in the fiber length has zero effect on the elastic modulus of the composite. The fiber particles are small in diameter. Although these fibers tend to have really good tensile properties, they can be bent easily when pushed axially. Hence, to prevent buckling and bending of the individual fibers, they must be constantly supported externally.

- Structural Composites consist of different layers of material with different orientations, held intact because of the matrix. Sandwich structures and Laminates are important examples in this category.

- Particulate Composites are made of a dispersed phase, which is in the form of particles, embedded in a matrix body. The dispersed phase particles may have preferred orientation (flakes) or have random orientation (powder form). Some examples that fall under this category include concrete and wood particle boards.

2.2 Types of polymer based matrix

The major functions of polymer based matrices in biocomposites are to permit the transference of tension between fibers, to prevent mechanical abrasion of surface fibers and to act as an effective barrier against hostile environments. The structural composites' tensile load carrying capacity is influenced by the function of the matrix. Key component of the biocomposite is the matrix or binding agent. Polymer, metallic, ceramic, and carbon are the various kinds of matrices generally used. Presently, polymer based matrices are highly utilized in industrial applications and these polymer resins are available in two kinds [6]:

a. Thermosetting: Upon heating application, the thermoset being a firm and inflexible cross linked component does not get molded [7]. They are pretty rigid and non-elastic in nature unlike elastomers and thermoplastics. Natural fiber composites use different kinds of polymers as matrices. Epoxy based resins such as fiberglass, phenols, polyamides, acids etc., are the frequently used thermoset polymers [3]. Due to their unique properties and their use in various applications, unsaturated polyesters are most preferred to be used as a matrix in a biocomposite. Owing to the polymer benefits such as room temperature curing property, decent mechanical characteristics, and clearness, they are manufactured more industrially in comparison to thermoplastic based resins. Studies have stated about cellulose fibers being reinforced with polyesters [8]. Few other systems which have similar potential are polyesters combined with pineapple leaf fibers, sisal, jute, coir, straw etc. [4].

b. Thermoplastics: These polymers tend to mold when heat is applied and can regain their shape after cooling. Their properties do not get affected even if they are reheated and reshaped numerous times. Polymethacrylate, high density polyethylene, polyvinyl chloride, low density polyethylene, polystyrene etc. remain the commonly used thermoplastics as matrix in biocomposites.

The kinds of thermoplastics that can be utilized for biocomposites are the ones whose processing temperature does not exceed above 230°C such as polyolefins, polypropylene, and polyethylene. Some of the technical based thermoplastics which need temperature > 250°C for processing cannot be utilized for creating biocomposites. The fibers have to be degraded initially for further use. Examples: polyesters, polyamides, and polycarbonates.

3. Natural fibers reinforced composite

Reinforced phase plays a very important role in determining the overall properties of the composite. Natural fibers can be broadly classified into 3 categories: Plant - based fibers, Animal - based fibers, Mineral - based fibers. One of the commonly used reinforcements is the mineral - based fibers, including carbon, fiberglass and Aramid [9].

3.1 Carbon fibers

Carbon fibers are unidirectional reinforcements. Due to this unique structural property of Carbon fibers, they can be structured in a way wherein the composite is stronger in a particular direction making it easier for the composite to bear heavy loads. The physical properties of carbon fiber can be modified by controlling different parameters such as the alignment of fiber, nature of the matrix, fiber-matrix volume fraction and the molding conditions.

3.2 Glass fibers

Glass is known to be extensively used as reinforcement in most of the polymeric matrix composites (PMCs). The main advantages observed during usage of glass fibers include low cost of production, good chemical resistance, high tensile strength, and excellent insulating properties.

3.3 Kevlar fibers

Kevlar fibers are known to belong to highly crystalline aramid (aromatic amide) fibers. These fibers have a very high ratio of tensile strength to weight and the lowest specific gravity among the currently used reinforcing fibers. Due to their superior mechanical properties, they tend to find major applications in marine and aerospace industries.

3.4 Boron fibers

Boron fibers are especially known for having extremely high tensile modulus. Another prominent feature of these fibers is the buckling resistance that results in a high compressive strength for boron fiber reinforced composites (**Figures 3** and **4**).

Though the present composite industry majorly depends on synthetic reinforcement fibers, the use of natural fibers have been gaining attention in recent years for academic as well as industrial purposes. In the present time, many different plant-based natural fibers have been explored and researched upon to identify unique properties. Some of these are used in plastics as an reinforcement; some examples of which include hemp, kapok, jute straw, paper mulberry, oil palm empty fruit bunch, wood, wheat, barley, kenaf, rye, rice husk, cane (sugar and bamboo), flax, reeds, oats, sisal, grass, coir, pennywort, water hyacinth, raphia, ramie, pineapple leaf fiber, banana fiber and papyrus [5].

One of the on-demand natural fiber reinforced composite is the thermoplastic matrix containing reinforcements made of special wood fillers due to them being light in weight, possessing reasonable stiffness and strength. Natural fibers have shown immense potential to be used in replacement to non-renewable materials due to their low cost, promising thermoplastic properties, minimal to zero health hazards and can act as a solution for environmental pollution [10, 12]. Some plant proteins have also been used as reinforcements. One such example is the wheat gluten, which when plasticized have a unique ability to form a strong cohesive

Figure 3.
Different types of natural fibers used as reinforcement in polymer composites.

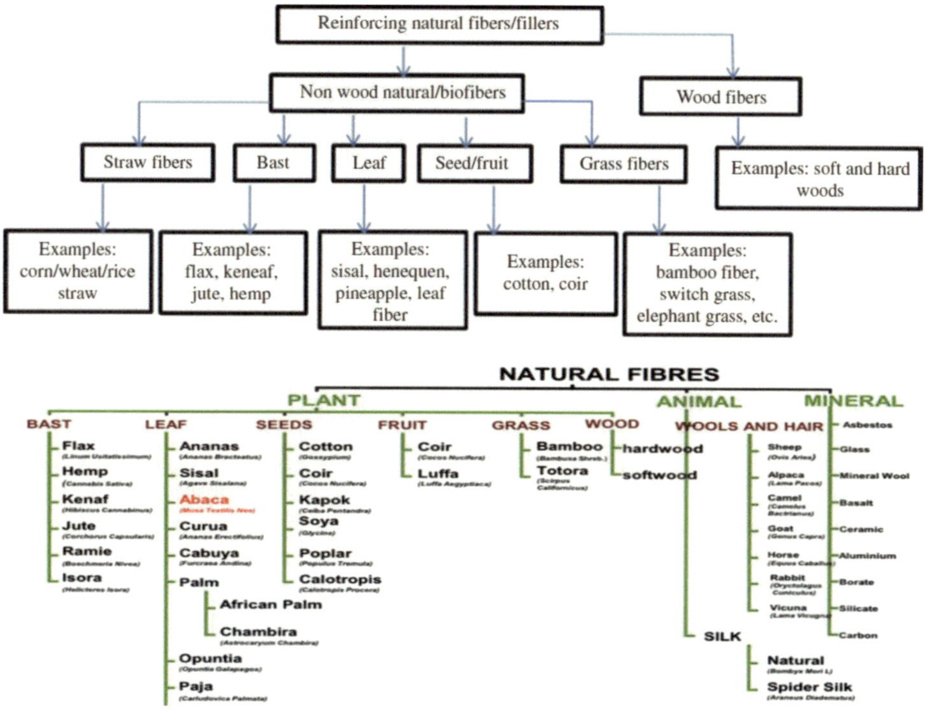

Figure 4.
Classification of various fibers of different origin and types with examples [10, 11].

blend with high viscoelastic properties [13–15]. Due to these distinctive properties, wheat gluten has been tremendously used in the making of packaging materials and edible biodegradable films.

Composites based on biologically degradable polyester amide and different plant fibers like flax and cottons, have been investigated thoroughly. These composites generally possess good mechanical properties, such as high biodegradability and good water resistance. Kenaf is a biodegradable and environmentally friendly crop belonging to the hibiscus family (*Hibiscus cannabinus* L). In a particular research work, Aziz et al. manufactured composite made of a polyester resin reinforced with Kenaf fiber and further studied the mechanical properties of the composite. The properties displayed by the composite were highly satisfactory and fiber could be blownto a height of at least 10 meter. By looking at the present research scenario, natural fiber reinforced polymer composites have the capability to act as a substitute for scarce wood and wood based materials that have many structural applications in the future [16] (**Figures 3** and **4**).

Plant - based fibers, also called vegetable fibers, are classified into different types based on their origin [17]. The characteristics of plant fibers majorly depend on certain factors such as the type of plant used, the area where it is grown, the plant's botanical age, and the protocol of extraction used. One such example is coir which is known to be a tough and hard fiber with multicellular layers, with the central portion called as "lacuna". Another familiar example is Sisal leaf fibers which are observed to have a high mechanical strength. Pineapple leaf fiber extracts are soft and are rich in cellulose. Oil palm fibers, having a similar cellular structure to coir, are hard and tough. Cellulose molecules make the major constituents in most of these plant fibers. The hydroxyls groups present in the basic unit of cellulose have the ability to form intra-molecular hydrogen bonds where the bonding is within the macromolecule, or intermolecular hydrogen bonding between two different cellulose macromolecules and or form hydrogen bonds with hydroxyl groups present in the atmosphere [18, 19]. It can be observed that all plant fibers have a high hydrophilicity, with their amount of retained moisture reaching about 8–13% [20]. Though cellulose is present in a huge quantity inside a plant, they also contain other natural substances such as lignin. The major role of lignin is to act as a cementing or bonding material between the cells of plant fibers. The content of lignin fibers influences a plant's structure, its morphology and its properties.

An important property of vegetable fiber influencing its reinforcement properties is the degree of polymerization (DP). The fibers differ drastically from each other due to the presence of cellulose molecules with differing DPs. Most of these fibers generally consist of a mixture of a base polymer homolog with the configuration $(C_6H_{10}O_5)_n$. The plant fiber known to illustrate the highest DP among other plant fibers is Bast fibers, with values nearing 10,000. In the olden times, these fibers found tremendous applications as packaging materials such as gunny bags and sacks, for making ropes, as a geo-textile material, for making twines and cords, and as carpet-backing [21, 22]. The most common bast fiber found in Cannabis sativa plants is Hemp, which is a lingo cellulosic fiber, repeatedly used as reinforcement in biodegradable composites. It is used in the making of various items such as shoes, toys and clothing, due to its non-toxicity, biodegradability and its ease for recycling [23].

3.5 Advantages of natural fiber reinforced composite

As mentioned earlier, natural fibers being used as reinforcement in composite materials tend to display good mechanical properties, thereby gaining a lot of attention in recent years. They are known to be fully biodegradable, renewable, environmentally friendly, cheap, available in abundance and have a low density [24]. Plant fibers are observed to be considerably light in weight when compared with mineral-based fibers. Due to their organic origins, plant fibers tend to possess high biodegradability which contributes to maintaining a healthy balance in the ecosystem. Due to the considerable low cost and high performance, these fibers can be termed as economically superior to other counterparts and can find huge applications in the industries. When natural fiber-reinforced plastics are combusted at the end of their life cycle, the amount of CO2 released from the combustion process is equal in amount assimilated during their growth [25]. Plant fibers have very low abrasivity which makes the recycling process of the composite materials much easier. Plastics reinforced with natural fibers, when used along with biodegradable polymers as matrix material, are considered to be the most environmentally friendly compounds because of their ability to be completely decomposed at the end of their life cycle [26, 27]. Natural fiber composites are used as a substitute to glass in non-structural applications. One such example is the automotive components that were initially manufactured with glass are now replaced with natural fiber reinforced composites [11]. Most of the plant fibers, when used in the unmodified form inside a composite, tend to produce unsatisfactory mechanical characteristics. To avoid this problem, the surface of plant fibers are treated with certain chemicals or compatibilizing agents prior to the fabrication of composite. The properties of these plant fibers can be further improved either by physical treatments such as corona treatment and cold plasma treatment, or by chemical treatments using peroxide, sodium hydroxide permanganate, isocyanates and maleic anhydride organosilanes [28]. In terms of mechanical properties of natural fibers, the values are quite low when compared to glass fibers or other mineral based fibers. When it comes to specific properties, especially stiffness, natural fibers values are near to those of glass fibers [29, 30]. Natural fibers have shown high tensile strength and stiffness values. Tensile strength within composites are majorly governed by the reinforcement used. Hence, natural fibers along with matrices can prove to produce the desired mechanical properties needed for the specific application [6, 9].

3.6 Disadvantages of natural fiber composites

Biocomposites function and quality majorly relies on biofibre characteristics. On the other hand, when biofibres are used for applications related to building materials, the composites have shown some disadvantages such as possessing high moisture absorption, low modulus elasticity, faster decomposition when surrounded by basic conditions, during microbial encounter, and visible variability in mechanical characteristics. In order to overcome these limitations, natural fibers need to be thoroughly studied to produce better performing fibers in future. Even though biofibers have various advantages for their usage in composites, they do possess certain disadvantages like being incompatible with other matrices, reduced wettability etc. [31, 32]. As mentioned briefly, the physical and chemical characteristics of biofibres are determined by cell wall based polymers and their matrix component (**Table 2**). The major properties of the fibers such as flammability, dimensional stability and biodegradability can be directly altered using external factors such as acids, bases, and UV rays by converting the green composite into carbon dioxide and water [30]. Therefore, to effectively overcome these limitations, there is a requirement to improve biofibre properties through altering the cell wall polymers chemistry [28].

In general, for achieving a decent function of short fiber reinforced polymer composites (SFRP), the content of fiber in the composite should be considerably higher [4]. The tensile characteristics of fiber reinforced is usually affected by occurrence of fiber and other materials used in the matrix [3]. Therefore, researchers are heavily focussed on studying the biofibers used in biocomposites and how they affect it [4].

Fiber	Density (g/cm^3)	Elongation (%)	Tensile Strength (MPa)	Elastic Modulus (GPa)
Hemp	1.47	2–4	550–1110	70
Banana	1.2	1.8	400	13–24
Ramie	1.47	3.6–3.8	400–938	61.4–128
Flax	1.4	2.7–3.2	500–1500	27.6
Kenaf	1.45	1.6	930	53
Jute	1.4	1.5–1.8	393–773	10–55
Sisal	1.3–1.5	2.0–2.5	511–635	9.4–22
Coir	1.2	30	287–800	4.0–6.0
Kraft	1.5	4.4	1000	40
Cotton	1.5–1.6	7.0–8.0	287–800	5.5–12.6
Bamboo	1.41	3.5	593	18–26
Kapok	1.69	0.7	100–1500	3–5
Phormium	1.2–1.4	0.9	250–310	26.5
Pine apple	1.1–1.3	1.2	188–308	11
Bagasse	1.2–1.3	1.3	300–350	12

Table 2.
Properties of selected natural and manmade fibers [6, 8, 9].

Natural fibers are generally known to be hydrophilic in nature and the polymer matrix tends to be more on the hydrophobic side. Due to this difference in their polarity, the composite material usually inclines to aggregate. There is increased water absorption as a result of hydrophilic fibers, showing a weak resistance to moisture. This further leads to producing poor tensile property reinforced composites. In addition to the basic cellulose unit, fiber surfaces contain wax coating and non-cellulosic compounds like lignin, pectin, and hemi-cellulose resulting in weak bonding among the biofibers and matrix.

Hydrophilicity being an undesirable property for obtaining natural fiber reinforced composite having good tensile properties, the fiber's hydrophobicity needs to be increased by using surface modifications (surface treatment). The natural fibers are modified for enhancing hydrophobic nature, roughness, interface bonding among the biofibres and matrix, and wettability. The modifications of the fiber also decrease the moisture absorption of the composite which gives us enhanced tensile properties [10, 12, 16, 17, 23].

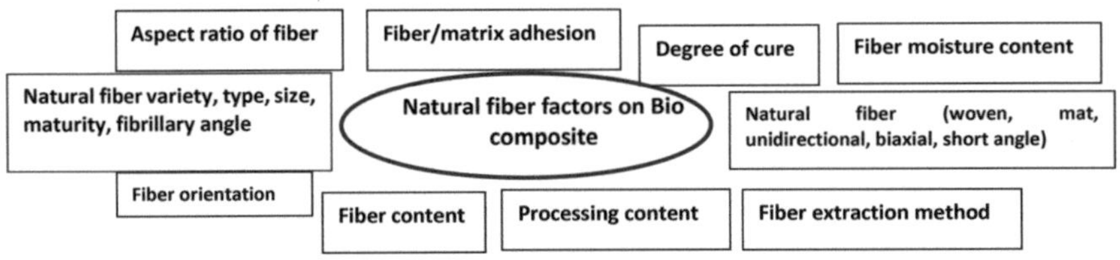

4. Applications of biodegradable polymers

i. Food Industry

The two main areas having applications in this sector are food packaging and edible films. The aim of food packaging is to enhance shelf life, ensure food safety, minimize food losses, improves the organoleptic properties of food like appearance, odor, and flavor. To replace the synthetic polymers, starch based biodegradable polymers could be utilized as an effective alternative with better properties. Although, the constituents involved in synthetic starch based packaging materials aren't fully inert. There is a possibility of migration of toxic substances into the food product which can affect humans. Due to this problem, alternative packaging materials are being studied [29, 32]. For example, investigations have led to the formation of starch or clay based nanocomposites that have shown low movement of polymeric substances, enhanced mechanical properties that can be utilized as a food packaging material. The characteristics of biodegradable films are colorless, flavorless, odorless, non-poisonous, and ecological. In low humidity conditions, these films exhibit very little penetrability to oxygen and help in increasing shelf life and quality of food product without compromising consumer acceptability. The compressed films or trays can be easily degraded by microbes as they readily dissolve in water medium. Therefore, starch based biodegradable polymers are of potential interest for food industry applications.

ii. Agriculture

In this sector, the key applications of biocomposites are greenhouse covering, mulch film, and fertilizer controlled release materials. Agricultural films are mainly consumed. Traditional films are disposed of by landfill, recycling or incineration. These processes take a lot of time, are not cost effective and pollute the environment. The important factor in developing agriculture productions is the effective usage efficiency of fertilizers. But, the fertilizers tend to escape to the environment because of surface runoff, leaching and vaporization thereby causing economic disadvantages and environmental issues. Starch based biodegradable polymers come into the picture to overcome these limitations [2]. The biocomposite can be utilized as a fertilizer controlled release matrix to release the fertilizer in the desired way. The films can be disposed of later after plowed into the soil. Due to their use, toxic residue formation does not take place after degradation. More studies are being explored involving starch based films in this sector. For instance, bio-nanocomposites are created by merging starch based film with additives like titanium dioxide, silicate or MMT to enhance mechanical characteristics.

iii. Medical industry

Starch based biocomposites can be utilized as an effective raw material due to their benefits such as biodegradability, biocompatibility, non-toxic, decent mechanical properties and degradation as needed.

Their usage in bone tissue engineering has been studied [22]. Structural framework support and degradation from the area of application is fastly offered by starch based biodegradable bone cements. Due to binning with bioactive particles, bone growth occurs at the bone cement interface and some amount resides because of polymer degradation. Bone tissue engineering scaffold involves the usage of these polymers. In drug delivery applications, starch based biopolymers are used. After drug depletion, this device does not have to be removed surgically. The starch based hydrogels or microspheres are used in different biomedical applications because of their novel properties like hydrophilicity, penetrability, biodegradability that mimics biological structures to some extent. Starch based biocomposites are of particular interest when it comes to biomedical applications [7, 11, 17].

4.1 A shift towards greener future

There is a lot of progress happening in the current field of research and development. Due to extensive usage of petroleum products, petroleum resources are getting exhausted and this also affects the environment and water bodies due to accumulation of plastics [7]. This is a huge concern as it affects the survival of human beings and animals. Because of these limited petroleum based resources, there is an urgent need to switch over to produce biodegradable alternatives. Biodegradable plastics are being produced to replace the synthetic polymers due to excessive usage of non-renewable polymer resources. Plants and crops can be used as suitable resources that can effectively replace the existing petroleum based products [11]. Biocomposites which are biopolymers reinforced with natural or biofibers can be utilized instead of synthetic composites such as glass fiber. Scientists are eyeing different approaches by merging biofibers like sisal, hemp, flax etc. with polymer matrix to produce biocomposites [11].

Composites are anisotropic in nature according to their structure. A matrix component (resin) is usually mixed with natural fibers obtained from plant based sources or cellulose to form biocomposites. Wood based fibers and non-wood fibers such as hardwood, softwood, jute, kenaf etc. are used as biofibres. The biocomposite's main constituent is the biofiber and it is obtained from tree, plant, or shrub sources [3]. The structure of biocomposites looks and functions similarly to living materials during the process and also enhances the strength properties of the matrix being used, thereby providing biocompatibility, for instance in forming scaffolds in bone tissue engineering. Structure and service environment are two important factors that determine the biodegradability rate in biopolymers. Natural/biofiber composites are suitable candidates as a potential replacement to the synthetic composites especially in various industries such as automotive, packaging, construction, and consumer products. It can also be used as an additive for thermoplastics. Additional investigations involve studying biological-inorganic interfaces to merge biological and inorganic materials while emphasizing on the design, production, and classification of novel amalgams [31]. Starch and cellulose are the most suitable renewable resources to create bioplastics. The cheapest source of biodegradable polymer in the marketplace today is starch and has applications in non-food industries as well. Cellulose plastics can be made by using cellulose from plant sources as an alternative to petroleum feed stocks [3, 7]. Vegetable oils have immense potential as a raw material to create biodegradable plastics, for example: plant oils and fats of sunflower, walnut, canola, sesame etc. Biocomposites created through biofibers and plant based bioplastics could be utilized in rigid packing, building, and transportation applications [3].

To manufacture products like decking, fencing, siding, window and door profiles etc., green composites are utilized. Some of the benefits obtained when biocomposites are utilized as construction materials are their low-cost, light in weight, eco-friendly, durability, and biorenewable [11].

4.2 Turning waste into bio composite

The bio composite was made using the waste materials. Billion-ton resource assessment conducted in 2011 has predicted sufficient supply of plant residues and

Figure 5.
Synthesis of ecofriendly biocomposite using the organic waste and analysis of water potential, elongation and tensile strength of composite material.

wastes within various price ranges. A supply of 180 million dry tons is expected at the end of 2030. When compared to primary plant residues, this supply seems low excluding animal manure. Corn stover, wheat straw accounts for at least three fourth of the residue resources whereas grains such as barley, sorghum, and wheat covers the rest. Availability of primary crop residues are enhanced by achieving continuous crop production and growth, large plots with less tillage, and no-till farming.

Green composites can be produced from different raw materials which are abundant in nature such as non-food crops and biorenewable sources. With recents trends in this particular field, biocomposite products could definitely find its place in the market across various applications. Different processing techniques and reinforcements of fibers/fillers allow the fiber at the correct price. Currently, there have been certain developments in the biopolymer industry.

In a study, the biocomposite ratio involving coconut blended with wheat husk was 15:85. Durability or toughness is affected by coconut husk due to it being fibrous in nature. Mixtures of wheat husk and water act as the matrix phase and binder. In terms of its mechanical properties, the product had displayed a strain of 26%, 7 N/mm^2 tensile strength along with a stair-step stress strain curve. The sample's performance might be because of various factors like how the axis is loaded, the relative motion among fibers through steady development across the loading axis. Crack proliferation due to interlinking of coconut husk has shown decent physical properties. A hybrid material is created by involving corn in the composite. Water absorption increases with increase in corn weight (**Figure 5**) [3].

5. Future prospects

Construction and automotive industry are the key markets for biocomposite materials. However, new prospects and applications will arise with certain future innovations and performance enhancement. Due to the accumulation of waste generated by this industry and their harmful effect on the environment which remains a growing concern, a lot of opportunities are going to emerge in this particular sector. Instances such as off-site construction method, for better quality and effortless installation and build, eco-friendly resources are required; although, these prospects could be affected by guidelines based on the present resources. A key target would be in replacing preservative-treated wood which provides a huge market growth. By placing strict limitations on utilization of preservatives such as arsenic containing products, provides immense opportunities for biocomposites to be used in applications especially when there is a threat of microbial attack.

Biocomposites can be incorporated into certain various complex technological applications by enhancing their mechanical functions such as producing new fiber types, processing methods, addition of additives etc. Solvent spinning process when applied to liquid crystalline cellulose creates high strength fibers which has been quite a hopeful research study. Resins can be formed by changing or enhancing the content of particular triglycerides and oils in produces by using biotechnology tools. The resins if altered appropriately would be cost effective and biodegradable compared to existing ones (**Figure 6**). Studies are being conducted to produce cheap biodegradable resins having decent mechanical characteristics by using novel methods. If successfully produced, the synthetic complexes can be replaced with these biodegradable alternatives.

Hybrid materials and products offer scope such as utilization of bio resins and bioplastics adhesives by replacing the existing fossil-based adhesives. Reclaimed fiber provides a decent opportunity to develop various eco-friendly, inexpensive

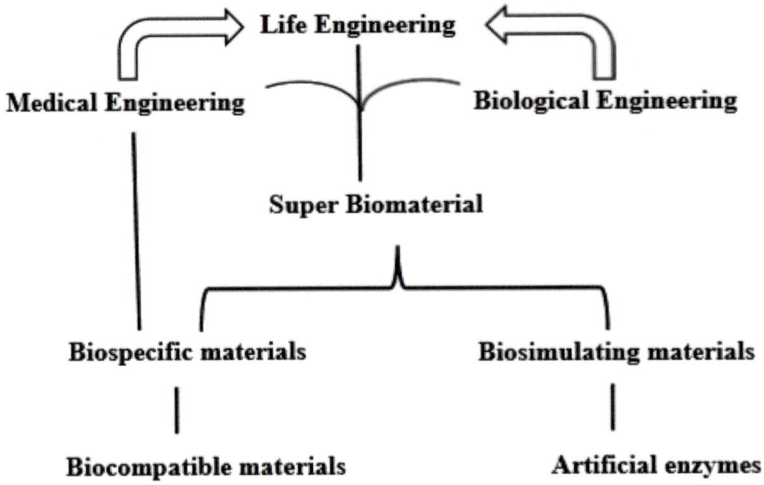

Figure 6.
Scheme for hypothesizing for engineering and biocomposites.

products by utilizing medium density fibreboards or the watercourses of the paper-making industry. Although sufficient prospects are there for these products to be in the market, cost effectiveness is a very important factor for its commercial production and hence the marketing strategy has to be made stronger accordingly. For successful commercialization, the biocomposites should be demonstrated through widespread training and education.

Author details

Kalmanje Mugdha Bhat, Jyothsana Rajagopalan, Rajeshwari Mallikarjunaiah, Nagashree Nagaraj Rao and Ashwani Sharma*
R.V. College of Engineering, Bengaluru, Karnataka, India

*Address all correspondence to: ashwanisharma@rvce.edu.in

IntechOpen

© 2021 The Author(s). Licensee IntechOpen. This chapter is distributed under the terms of the Creative Commons Attribution License (http://creativecommons.org/licenses/by/3.0), which permits unrestricted use, distribution, and reproduction in any medium, provided the original work is properly cited.

References

[1] M C Khoathane, O C Vorster,and E.R Sadiku, Hemp Fiber-Reinforced 1-Pentene/Polypropylene Copolymer: "The Effect of Fiber Loading on the Mechanical and Thermal Characteristics of the Composites, Journal of Reinforced Plastics and Composites", Vol. 27, 2008, pp.1533–1544.

[2] R Malkapuram, V Kumar and S N Yuvraj, "Recent Development in Natural Fiber Reinforced Polypropylene Composites", Journal of Reinforced Plastics and Composites, Vol. 28, 2008 , pp. 1169–1189.

[3] Ashwani Sharma, Nagashree N Rao and M S Krupashankara. Development of Eco-friendly and biodegradable Bio composites. Materials Today Proceedings, 5, 2018 pp, 20987–20995.

[4] P Wambua, J Ivens, and I Verpoest, Natural Fibers: "Can They Replace Glass in Fiber Reinforced Plastics", Composites Science and Technology, Vol. 63, 2003 pp.1259–1264.

[5] J Holbery, and D Houston, "Natural-Fiber-Reinforced Polymer Composites in Automotive Applications", JOM, Vol. 58, No. 11, 2006 pp. 80–86.

[6] D Nabi Saheb, and Jog, J P, Natural Fiber Polymer Composites: A Review, Advanced in Polymer Technology, Vol. 18, 1999 pp. 351–363.

[7] K Mugdha Bhat , Jyothsana R, Ashwani Sharma, Nagashree N Rao, Carrageenan-based edible biodegradable food packaging: A review, International Journal of Food Science and Nutrition Vol 5 (4), 2020 pp 69–75.

[8] Li, X., Tabil, L G, S Panigrahi, and W J Crerar, "The Influence of Fiber Content on Properties of Injection Molded Flax Fiber-HDPE Biocomposites", Canadian Biosystems Engineering, 08–148, 2009 pp. 1–10.

[9] I Ahmad , A Baharum, and I Abdullah, "Effect of Extrusion Rate and Fiber Loading on Mechanical Properties of Twaron Fiber-thermoplastic Natural Rubber (TPNR) composites", Journal of Reinforced Plastics and Composites, Vol. 25, 2006 pp. 957–965.

[10] B H Lee, H J Kim, and W R Yu, "Fabrication of Long and Discontinuous Natural Fiber Reinforced Polypropylene Biocomposites and Their Mechanical Properties", Fiber and Polymers, Vol. 10, 2009 pp. 83–90.

[11] Kurki Nagaraj Bharath and Satyappa Basavarajappa, "Applications of biocomposite materials based on natural fibers from renewable resources: a review", Sci Eng Compos Mater 2016; 23(2): 123–133

[12] X Li, S Panigrahi, and L G Tabil, "A Study on Flax Fiber-Reinforced Polyethylene Biocomposites, Applied Engineering in Agriculture", Vol. 25, 2009 pp. 525–531.

[13] H Hajnalka, I Racz, and R D Anandjiwala, "Development of HEMP Fiber Reinforced Polypropylene Composites", Journal of Thermoplastic Composite Materials, Vol. 21, pp.165–174, 2008.

[14] A.G Facca, M.T Kortschot, and N Yan, "Predicting the tensile strength of natural fiber reinforced thermoplastics", Composites Science and Technology, Vol. 67, 2007 pp. 2454–2466.

[15] A.G Facca , M.T Kortschot, and N Yan , "Predicting the elastic modulus of natural fiber reinforced thermoplastics, Composites": Part A: Applied Science and Manufacturing, Vol. 37, 2007 pp.1660–1671.

[16] B.S Panigrahy, A Rana, , P. Chang, and S Panigrahi, "Overview of Flax Fiber Reinforced Thermoplastic

Composites", Canadian Biosystems Engineering Journal, 2006, 06–165, pp. 1–12.

[17] E F Santos, R S Mauler, and S M B Nachtigall, "Effectiveness of Maleated- and Silanized-PP for Coir Fiber-Filled Composites", Journal of Reinforced Plastics and composites, Vol. 28, 2009 pp. 2119–2129.

[18] R Hu, and J K Lim, "Fabrication and Mechanical Properties of Completely Biodegradable Hemp Fiber Reinforced Polylactic Acid Composites", Journal of Composite Materials, Vol. 41, 2007 pp.1655–1669.

[19] M A Fuqua, and C A Ulven, "Preparation and Characterization of Polypropylene Composites Reinforced with Modified Lignocellulosic Corn Fiber", The Canadian Society for Bioengineering, (08) 4364, 2008.

[20] L Liu, J Yu, L Cheng, and W Qu, "Mechanical Properties of poly (butylenes succinate) (PBS) biocomposites reinforced with surface modified jute fiber", Composites Part A, Vol. 40, 2009 pp. 669–674.

[21] P J Herrera-Franco, and A Valadez-Gonzalez, "Mechanical Properties of Continuous Natural Fiber-Reinforced Polymer Composites", Composites Part A: Applied Science and Manufacturing, Vol. 35, 2004 pp. 339–345.

[22] S.Tungjitpornkull, and N Sombatsompop, "Processing Technique and fiber orientation angle affecting the mechanical properties of E-glass reinforced wood/PVC composites", Journal of Materials Processing Technology, Vol. 209, 2009 pp.3079–3088.

[23] X. Ma , J.Yu, and J.F Kennedy, "Studies on the Propertied of Natural Fibers-Reinforced Thermoplastic Starch Composites", Carbohydrate Polymers, Vol. 62, 2005 pp. 19–24.

[24] S. Husseinsyah and M. Mostapha, "The Effect of Filler Content on Properties of Coconut Shell Filled Polyester Composites," *Malasian Polymer Journal*, Vol. 6, No. 1, 2011 pp. 87–97.

[25] X Li, L G Tabil , S Panigrahi, and L;. G Tabil, "A Study on Flax Fiber-Reinforced Polyethylene Biocomposites by Injection Moulding", Applied Engineering in Agriculture, Vol. 25, 2009 pp. 525–531.

[26] L Medina, R Schledjewski, and A K Schlarb, "Process Related Mechanical Properties of Press Molded Natural Fiber Reinforced Polymers", Composites Science and Technology, Vol. 69, 2009 pp.1404–141.

[27] O A Khondker, U S Ishiaku, A Nakai, and H Hamada, "Fabrication and Mechanical Properties of Unidirectional Jute/PP Composites Using Jute Yarns by Film Stacking Method", Journal of Polymers and Environment, Vol. 13, 2005 pp. 115–126.

[28] S N Monteiro, L A H Terrones, and J R M D'Almeida, Mechanical Performance of Coir Fiber/Polyester Com-posites, Polymer Testing, Vol. 27 (5), 2008 pp. 591-595.

[29] W Wang, and G Huang, Characterization and Utilization of Natural Coconut Fibers Composites, Materials & Design, Vol. 30 (7), 2009 pp. 2741-2744.

[30] M N Angles, and A Dufresne, Plasticized / tunicin whiskers nanocomposites materials, Mechanical properties. Macromolecules, Vol.34 (9), 2001 pp. 2921-2931.

[31] Curtu and D Motoc Luca, Theoretical and Experimental Approach of multiphase composite materials, B. Katalinic, Ed. Vienna: DAAAM International Publishing, 2009 pp. 349-362.

[32] I Z Bujang, M K Awang, A E Islmail, Regional conference on Engineering Mathematics, Mechanics, manufacturing & Architecture. 185-201, 2007.

Chapter 2

Opportunity of Non-Wood Forest Products in Biocomposites

Pradeep Sharma

Abstract

In recent years industries are attempting to decrease their dependency on petroleum-based fuels and products due to increased environmental issues. The tremendous increase in production and use of plastics in every sector of life has led to huge plastic waste disposal problems and also an environmental threat. In order to prevail over the present scenario, the viable and cost-effective approaches are to prepare eco-friendly bio-composites based on non-wood forest products (NWFP), a part of forest wealth of the globe, especially natural fibres, agricultural wastes and extractives. Natural fibres and extractives have many advantages viz. low density, low cost, considerable toughness properties, nontoxicity, sustainability and bio-degradability. NWFP based composites may be utilized to produce non-structural parts for diverse applications in various industries as high-performance materials with interesting properties for specific applications viz. furniture, thermal, acoustic insulations and automotive industries etc. In the present chapter, opportunities of extractives, cellulosic and lignocellulosic fibres from non-wood forest products in Bio-composites will be discussed.

Keywords: Non-wood forest products, Bio-composites, Tannins, Laccase, Fibres, Agriculture wastes

1. Introduction

Biocomposites are composites formed by the mixture of two or more than two constituents which are firmly stacked in a specific orientation in order to provide stability and toughness as per their requirement. Composites may be synthetic, biocomposites or natural composites. Natural composites are the wood, bamboo, silk, cellulose, and some animal products viz. feather, horn etc. Bio-composites are materials formed by reinforcement of natural fibres into adhesive or a matrix. The matrix may be a natural, synthetic material or an amalgamation of natural and synthetic materials. Environmental concerns over the synthetic matrix and further cost of synthetic fibres have led the encouragement of scientific community of using natural fibres as reinforcement material in polymeric composites.

2. Non wood forest products

The non wood forest products (nwfps) comprise all the forest products other than timber and fuelwood and are used by human beings since the time immemorial [1–3]. NWFPs include medicinal plants, essential oils, spices, edible wild

IntechOpen

plants, gums, resins and oleoresins, fatty oils, tanning materials, natural organic colouring materials, katha and cutch, oxalic acid, fibres and flosses, beverages and narcotics, fodder and forage plants, saponins, fish poisons, insecticides, green manure, beads, rubber plants, plants useful for paper, basket and wicker work including canes, beedi leaf etc., miscellaneous materials including thatching and broom materials. Besides these plant products, animal products such as lac, honey, silk, horns, ivory and hides (of forest origin) are included among the nwfps [4–6]. Developing countries especially tropical region of the world more than three-fourths of the populations are dependent on nwfps for their nutrition, primary health care and livelihood subsistence. Therefore, nwfps play an important role in the daily lives of local population in particular rural and poor people dependency on nwfps for their daily needs of food, fodder, medicines, gums, construction material, etc. In addition to local consumption, nwfps are also traded in local, regional, national and international markets and the trade in nwfps not only generates employment opportunities but also contributes in the economic development of the country [7–9]. Among the nwfps fibres and flosses, bamboo and canes, tans and dyes, essential oils are important forest bio-products for livelihood support to marginal peoples residing in forest areas. After processing of the essential products (dyes, essential oils etc.), the left over biomass may be utilized for diverse industrial applications. The importance of these nwfps lies in the following facts.

2.1 Fibres and flosses

A wide range of plants yielding fibre occur in the forests wealth. Fibres are obtained from tissues of different parts of certain woody plants, which are used for various traditional applications such as making cloth, rope, mat and cordages etc. [10, 11].

2.2 Bamboo

Bamboo is abundantly found in most parts of the world, nearly 0.92 % of the total forest area, spread over 36 million hectares (MHa) [12]. Globally, bamboo has 111 genera with more than 1575 species. India is very rich in bamboo resources and the second major bamboo producing country having 16 MHa (22.46 %) of a total forest cover 71.2 MHa [13] comprising of 160 species after China. China is the richest in bamboo resources; it has more than 800 species [14, 15]. The bamboo is used for various industrial purposes [16–18]. Bamboo species thrives in almost all types of soil except in very dry soils. The bamboo is utilized for various purposes, viz. agriculture, handicraft, building industry (bamboo concrete, scaffoldings, house construction, etc.), interior decoration (Bamboo flooring board, mat, panelling, curtain, etc.), paper industry, textile industry, food, bamboo charcoal, and diverse range of daily use articles (toothpicks, chopsticks, incense sticks, etc.) [19–22]. Bamboo is a typical natural composite material with functionally gradient structure having multi-nodes, and the fibres are arranged compactly in the outer surface region in a definite fashion in comparison to inner surface region which provides fracture toughness. The fracture toughness of the bamboo culm depends on the volume fraction of fibres [23–28].

2.3 Essential oils

Essential oils from plants are widely used in pharmaceuticals, cosmetics, perfumery and specialty applications in various industries. The value of essential oil bearing crops can be augmented by utilization of wastes using facile and economic methods. The major essential oil bearing crops are lemon grass, mentha, eucalyptus, lavender, rose, geranium, rosemary, basil, thyme, peppermint, chamomile, etc. [5].

2.4 Tans and dyes

A variety of vegetable tanning materials are produced in the forests. Important vegetable tanning materials are the myrobalan nuts and bark of wattle (*Acacia mearnsii*, *A. decurrens*, *A. nilotica* and *Cassia auriculata*, etc.). Other tanning materials include leaves of *Emblica officinalis* and *Anogeissus latifolia,* bark of *Cleistanthus collinus*, *fruits of Zizyphus xylocarpa*, bark of *Cassia fistula*, *Terminalia alata*, *T. arjuna*, etc. The term tannin was introduced by Seguin in the year 1796 to indicate various plant extracts, which have the capacity to convert hides and skins into leather [5, 29, 30]. Tans and dyes are simple chemical compounds made of carbon, hydrogen and oxygen. The structure and solubility of tannins are dependent upon the source and structure of tannins, however, vegetable tannins are water-soluble.

2.5 Wood tans

The important species yielding wood tans are Quebracho (*Quebracho Colorado*), this species is widely distributed in South America. The heart-wood contains 20-27 % tannin, which is obtained by cutting the wood into small chips and extracting the tannin with water. The cutch, a byproduct from Katha industry, obtained from khair (*Acacia catechu*) heart wood is used for tanning purposes.

2.6 Bark tans

The bark of various tree species is chipped off during the operation of timber or fuel-wood harvesting, some of the important species which yield bark tans are: *Acacia mearnsii, Acacia nilotica, Cassia auriculata, Shorea robusta, Terminalia arjuna, Cassia fistula, Ceriops roxburghiana, Rhizophora mucronata,* etc. Some other trees such as *Acacia leucophloea, Bridelia retusa, Lagerstroemia* spp., *Tamarix aphylla, Terminalia alata, Quercus* spp. and *Castanopsis* spp. also yield bark tans and are locally important.

2.7 Fruit tans

Fruits of some of the forest trees are utilized in tanning industries for extraction of different tannins. Some of the important species yielding fruit tans are as follows: *Acacia nilotica* (Babul), *Caesalpinia coriaria* (Divi-divi), *Zizyphus xylopyrus* (Kath bor), *Emblica officinalis* (Aonla), *Shorea robusta* (Sal), *Anacardium occidentale* (Cashew nut), *Tamarindus indica* (Tamarind) and *Terminalia chebula* (Myrobalan), etc.

2.8 Leaf tans

Leaves of some of plants provide tanning material however; they are not used on a large scale for commercial applications. Generally, village artisans and shoemakers use leaves for tanning leather on as a small scale. Important leaf tanning materials are obtained from leaves of *Anogeissus latifolia, Carissa spinarum, Emblica officinalis, Lawsonia inermis* and *Rhus cotinus,* etc.

2.9 Natural dyes

Natural dyes are broadly classified as plant, animal, mineral, and microbial dyes. They are obtained from the vegetable plant materials such as plant leaves, roots, bark, seeds, and from insect secretions and minerals. Dyes are used for various industrial applications including the coloring of textiles [31, 32].

3. Natural composites

Wood and bamboo are the natural composites, held together by the matrix as designed by the nature (**Figure 1**). Lignin is the largest biopolymer and principle cementing matrix which holds the components of natural composites together in a definite fashion. The awareness in the society globally and harmful effects of the synthetic materials on the environment has led to the progressive development of eco-friendly and sustainable materials. The scientific community have shown a lot of interest in developing sustainable bio-composites which are eco-friendly and may substitute partly or wholly the synthetic materials. Intriguingly, using natural fibres in development of biocomposites provides a reduction in greenhouse gas emissions and carbon footprint of composites [33, 34]. Due to environmental concern, demand for commercial raw material for utilization in composites increasing day by day. Therefore, after harvesting the important chemicals or fresh materials from the nwfps may be utilized on sustainable basis for making bio-composites for various industrial sectors. Among them, use of the natural fibres as reinforcement material, tannins and lignin as cementing material alone, in parts or in combination with synthetic matrix subsequent to proper modification, and further, valorisation

Figure 1.
Natural Wood Composite.

DOI: http://dx.doi.org/10.5772/intechopen.97825

of residues from the nwfps is an important opportunity as raw materials for bio composites in a sustainable way.

4. Natural fibres and their utilization in bio-composites

A vast number of research papers and reviews are available for utilization of natural fibres in sustainable development of biocomposites. However, as mentioned, the fibres are sustainable raw materials and may be obtained from different sources from nwfps, further, low-cost, light-weight, availability, renewability, biodegradability; properties and strength are important factors for their utilization. The specific properties of the fibres are utmost important in developing the composites. Natural fibres comprise of cellulose, hemicelluloses, lignin, waxes and tannins etc. The percentage of these constituents varies with the source and processing of fibres. Further, the properties of biocomposites are also dependent upon the source of fibres and presence of these constituents. Natural fibres are obtained from plants or animals [35]. The plant fibres are commonly used for producing bio-composites and mostly sourced from nwfps or in some countries cultivation of these crops is being carried out for sustainable production. The commercial sources of fibres utilized in producing biocomposites worldwide due to their inherent properties. The classification of fibres based on source, their specific properties and utilization in biocomposites are summarised in **Tables 1-3**.

Natural fibres show a variation in properties. The fibre properties are dependent on the geographical location, process of isolation (ratting of fibres) and maturation period. However, some fibres exhibit highest tensile strength in a range from 300-1100 MPa (**Table 1**).

4.1 Merits and demerits of natural fibres

The inherent properties of the natural fibres of plants origin are important in developing the bio-composites. Natural fibres comprise of cellulose, hemicelluloses, lignin, waxes and tannins etc. The percentage of the cellulose, hemicelluloses, lignin etc., length and width of the fibres, varies with the source and processing of fibres. Further, natural fibres possess low density (1.25–1.50 g/cm^3), sufficient mechanical properties, sustainability, recyclability, biodegradability, availability and low-priced in comparison to synthetic fibres such as glass and carbon fibres [51, 52]. Intriguingly, these properties do not meet the requirements of biocomposites. The natural fibres are used for increasing the mechanical strength as reinforcement material in composites [53]. The synthetic matrix and natural fibres are not compatible to each other leading to poor mechanical properties properties. Further, fibres have also water absorption capacity of cellulose due the presence of numerous hydroxyl groups [54–56].

Natural fibres are used as reinforcement materials in composites. However, due to their susceptibility to moisture [56] mechanical properties of polymeric composites have a strong impact on the interface adhesion between the fiber and the polymer matrix [54]. The natural fibres are rich of cellulose, hemicelluloses, lignin, pectins, waxes and tannins etc, all of which are composed of hydroxyl groups. Thus, there are major challenges of suitability between the matrix and fiber that weakens interface region between matrices and natural fibres [55]. Generally, outer surface of the composite materials absorb water and decreases gradually into the bulk of the matrix. High water absorption capacity of the composite materials leads to decline in their mechanical strength and pressure on nearby structures due to absorption of water pertaining to the hygroscopic nature of the fibres and

Fibres Source	Common name	Part of the plant	Appearence	Diameter (μm)	Density (g/cm^3)	Tensile strength (MPa)	Young' modulus (GPa)
Agave sislana	Sisal	Leaf	Coarse-stiff; Creamy-White	50-200	1.45	468-640	9.4-22
Corhorus capsularis	Jute fibre	Bast	Fine Light brown	25-200	1.3-1.45	393-773	13-26.5
Cocos nucifera	Coconut fibre or Coir	Coconut fruit outer shell or husk	Course, white to brown	100-450	1.15	131-175	4-6
Arenga pinnata	Sugar-palm fibre	Bast, fruit, leaf	Coarse, brownish-black	50-800	1.29	190.29	3.69
Cannabis sativa	Hemp fibre	Bast	Silky-fine, white to light brown	26.5	1.48	514	24.8
Bamboo Species	Bamboo	Grass	Creamy white	10-40		503	35.91
Linum usitatissimum	Flax fibre	Bast	Creamy white	12-16	1.50	345-1100	27.6
Hibiscus cannabinus	Kneaf fibre	Stem	Pale	17.7-21.9	1.2-1.4		
Boehmeria nivea	Ramie fibre	Bast	White	25-30	1.5-1.55	400-1600	44
Gossypium Sp.	Cotton fibre	Seeds	Yellowish off-white	11-22	1.51	400	12
Conifers	Soft wood fibre	Soft wood tracheids	Variation in colour	20-35	1.5	1000	40
Ananas comosus	Pine apple	Leaf	White, smooth and glossy	8.66-63.43	1.44	413-1627	34.5-82.5

Table 1.
Plant fibres and properties [36–45].

subsequently can cause warping, buckling, bigger possibility of their microbial inhabitation, freeze, and unfreeze leading to destruction of mechanical characteristics of composite materials. Therefore, fibres are required to improve these limitations by physical and chemical modifications [57].

4.2 Alteration of properties of natural fibres

The compatibility of the natural plant fibres with the synthetic matrix is the main and foremost concern of developing bio-composites due to the different nature and properties of these two materials. The various methods have been studied and reviewed in the past in order to increase the functionality and compatibility of natural fibres. Fibres compatibility with the matrix and mechanical strength thereof may be increased by physical and chemical modification of the fibres.

Fibre	Cellulose (%)	Hemicelluloses (%)	Lignin (%)	Pectin (%)
Cotton	82–96	2–6	0.5–1	5–7
Pine apple	80–81	16–19	4.6–12	2–3
Hemp	70–92	18–22	3–5	0.9
Flax	72–84	16–18	0.6–5	1.5
Ramie	68–76	13–15	0.6–1	1.9–2
Abaca	61-64	21	12	0.8
Nettle	86	5.4	4	-
Banana	60–65	6–19	5–10	3–5
Jute	51–84	12–20	5–13	0.2
Kenaf	49-53 (alpha)	86.8-87.7 (Holocellulose)	14.7-21.2	-
Lemom grass	71.7	9.52	13.83	-
Coir	46	0.3	45	4
Sisal	43–78	10–13	4–12	0.8–2
Bagasse	32-48	21	19.9-24	10
Bamboo	26-43	15-26	21-33	-
Kapok	53.40-62.9	29.63	19.2-20.73	-
Henequen	77.6	13.1	4.8	-
Hard wood	43-47	16-24	25-35	-
Soft wood	40-44	25-31	25-29	

Table 2.
Chemistries of important plant fibers [43, 46–50].

Type of Fibres	Source	Plant of fibres	Use	Properties
Bast fibres	Stem	Jute	Reinforcements	Long length high stiffness and strength
	Stem	Ramie	Reinforcements	Long length high stiffness and strength
Stalk	Bamboo	Bamboo species	Reinforcement	Short and long fibres
	Wood	Soft and hard wood	Reinforcement	Short fibres
Leaf	Leaf	Banana	Reinforcement	Long fibres
	Leaf	and sisal	Reinforcement	Long fibres
Seeds	Cotton	Cotton species	Reinforcement	Short fibres
	Agricultural wastes	Cereals etc.	Reinforcement	Short and long fibres

Table 3.
Plant based fibres, source and utilization in biocomposites.

4.2.1 Physical modification

The surface properties may be increased by physical treatments of the fibres. However, during extraction process, the journey of fibres to a final destination also involves the multi stepping process leading to stress and physical changes in the inherent properties of fibres. During the extraction process there are some fibres which involve simple process of extraction for example *Agave sislana* fibres. In most of the cases, the processing of plant material containing raw material for bio-composites as fibres encompass the physical and mild chemical treatments leading to change in original properties of fibres. Therefore, extraction of fibres is also an important factor in considering the evaluation of properties of bio-composites. There are various processes developed and optimized for extraction of fibres and well documented.

The physical treatments of the isolated fibres change the structural and surface properties of the reinforcing fibres without altering the properties and disintegration of fibres. The physical treatments influence the mechanical properties resulting in proper bonding to the matrix and affects interfacial adhesion. The commonly used method for plant-based fibres is corona and cold plasma treatment, however other physical methods are also successfully used for surface activation such as thermotreatment [58, 59]; calandering [60], stretching [61] and hybrid yarns [62]. The corona treatment provides oxidation of the fibres, which changes the surface energy of the fibres and increases the number of aldehyde groups [63–65]. The corona and cold plasma treatment are called electric discharge methods and mostly used to activate cellulose fibres leading to increase in mechanical strength [63, 64, 66].

4.2.1.1 Corona treatment

Corona treatment is employed for treatment of fibres to increase the morphological and mechanical properties of lignocellulosic fibres resulting in an improvement of the interfacial compatibility between matrix and fillers. Homogeneity of composite materials, adhesion properties and mechanical properties (tensile strength, Young modulii) increase to a certain level (10–30%) with corona treatment [67, 68]. Recently, *Aloe vera* fibres [69] were treated with corona discharge during different time intervals and it was observed that rough surface morphology and degradation of fibres occurred due to etching mechanism caused by corona treatment.

4.2.1.2 Plasma treatment

Plasma treatment is an environmentally friendly green electric discharge method for treatment of fibres [70–72] and provides changes in surface energy, increase of the roughness and micro-cleaning of the treated fibres. The process causes surface crosslinking and can introduce reactive groups. Mostly low plasma treatment is being carried out in presence of gases to alter the surface properties of fibres. The base material is treated under atmospheric plasma glow discharge for various periods of time using helium, helium/nitrogen, and helium/acetylene, argon, oxygen, air etc. gas. The significance lies in the fact that sometimes desired properties obtained in seconds. Intriguingly, changes in surface roughness, tip-surface adhesion, and surface chemistry of the fibres and flexural strength, flexural modulus, and interlaminar shear stress, storage modulus and glass transition temperature increased significantly. The treatment is successfully employed to alter the surface properties of natural fibres used in composites as reinforcing material. The adhesion between sisal fibres and polypropylene matrix [73] increase the interfacial adhesion between flax fibre and matrix polyethylene and unsaturated polyester [74, 75]; improvement in mechanical properties of ramie fibres [76, 77] polypropylene composites, increase in flexural strength and tensile strength of the composite prepared from jute fibre [78–81] were obtained by employing plasma treatments.

4.2.2 Chemical modification of fibres

Natural fibres possesses high polarity in nature due to the presence of numerous hydroxyl groups on the fibre surface which makes them incompatible with the synthetic hydrophobic matrix resulting in poor interfacial bonding between the cellulosic fibres and the matrix producing bio-composites with lesser physical and mechanical strength. Chemical treatments of fibres are an important step for processing of bio-composites and enhancing the compatibility of fibres to the synthetic matrix. Bi-functional groups are introduced chemically into the fibres leading to activation of hydroxyl groups. The activated hydroxyl groups further react with the synthetic matrix thereby enhancing the interfacial adhesion and compatibility between the fibres and matrix [76, 82]. Reviews and chemical methods have been reported in the past to increase the functional behaviour of hydroxyl groups present in the polysaccharides. These methods have their own limitations for particular industrial aspects and are used for diverse industrial applications [83–94]. Mercerization, benzoylation, acrylation, acrylonitrile grafting, permanganate treatment, peroxide treatment, and isocyanate treatment, bleaching, acetylation, silane and peroxide treatments, etherifications viz. cynoethylation, quaternization, carboxymethylation and various coupling agents are commonly used for lignocellulosic fiber activation [76, 87]. Mercerization of fibres using alkali at different concentration and time period for activation and production of modified fibres with good adhesion properties have been applied for modification of fibres [95, 96]. Transverse strength of flax fibres may be increased sufficiently by the alkaline treatment which led to produce better adhesion properties between flax fibres and epoxy matrix [97]. Introduction of acetyl groups into the cellulosic fibres increases plasticity leading to hydrophobic character and mechanical strength to the reinforcing material [76, 98]. Coupling agents are frequently and successfully used to reduce the interfacial adhesion of fibres to the matrix. Various organosilanes mostly trialkoxysilanes are variably used as coupling agents and the process is referred as Silanization. The reactive alkoxy groups present in the silanes chemically bond with the hydroxyl groups and the formation of polysiloxane structures occurs [58, 99]. Maleic anhydride is another coupling reagent used to increase the interfacial adhesion of biocomposites [100]. Partial removal of lignin on the henequen fibres increases the adsorption of the silane couplings and interaction among the fibre and the matrix [101]. The mechanical properties including tensile, flexural, impact strengths and tensile modulus of the biocomposites were improved several times on Jute fibre polypropylene composites using m-isopropenyl-α-α-dimethylbenzyl-isocyanate (m-TMI) as the coupling agent using grafting process. Further, the tensile modulus of the composites prepared from virgin polypropylene increases manifold [102]. The use of the clay in the bio-composites formulation led to reduced mechanical properties. Intriguingly, techniques such as pre-coated fibres with nanoclay and maleated polyethylene mixture enhance the synergetic effect of the clay and bamboo fiber and further significantly increase the tensile strength, bending modulus and strength of the high density polyethylene bamboo composites [103].

Maleated coupling agents are widely used to strengthen composites containing fillers and fibre reinforcements. The maleated coupling provides efficient interaction of maleic anhydride with the functional surface of fibres and matrix. Agrofibre polypropylene composites were studied by introducing maleated coupler that provides the flexural and tensile strengths by more than 60% with Epolene™ G-3015 increment in comparison to composites without coupler [89]. Maleic anhydride grafted rice husk [104], hemp fibres unsaturated polyester composites coupled with 3-isopropenyl-dimethylbenzyl isocyanate [105], maleic anhydride-grafted polypropylene jute fibre composites [106], coir fiber and m-isopropenyl-α-α-dimethylbenzyl isocyanate grafted polypropylene composites [107] may be implemented in production of superior biocomposites having high mechanical properties and strength.

5. Biocomposites from extractives

5.1 Biocomposites using tannins

Tannins are a group of polyhydroxy phenolic compounds and exhibit good alternatives to synthetic adhesives for green chemistry in developing composites. They are found abundantly in nature. Their functions are to protect the plants against predation and might help in regulating the plant growth. Tannins are heterogeneous in nature and chemically classified into two main groups viz. hydrolysable and condensed tannins. Hydrolysable tannins are small molecular weight (30-3000D) compounds, heterogeneous in nature and hydrolysed by water, acidic or alkaline conditions into smaller water soluble molecules such as gallic acid and ellagic acid (**Figure 2a** and **b**) constituting gallotannins and ellagitannins. The gallotannins and ellagitannins comprise of a central sugar unit esterified with several molecules of

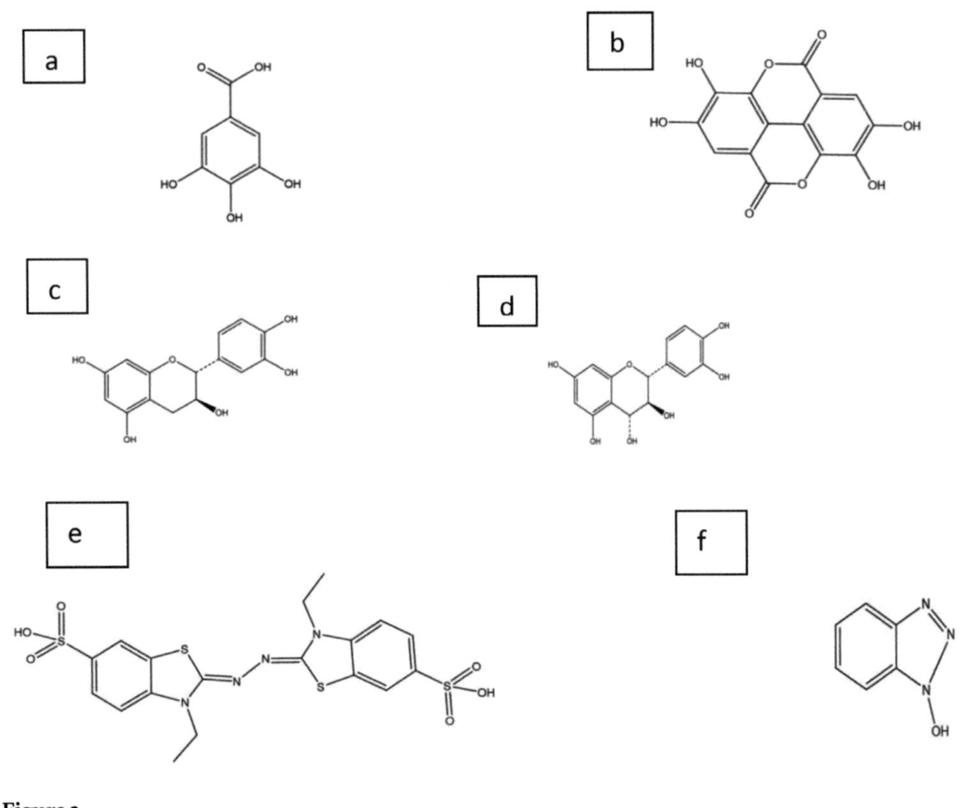

Figure 2.
(a) Gallic acid (b) Ellagic acid (c) 5,7,3,4- tetrahydroxyflavan– 3– ol (d) 5,7,3,4- tetrahydroxyflavan– 3,4-diol (e) ABTS (2,2'-azino-bis(3-ethylbenzothiazoline-6-sulfonic acid) (f) HBT (1-hydroxybenzotriazole).

gallic acid and a dimer of gallic acid as the basic phenolic unit known as ellagic acid respectively. Gallotannins, or commonly tannic acid, is the acknowledged source of the hydrolyzable tannins produced by extraction with water or organic solvents from the galls of certain trees, *Quercus infectoria* and *Rhus chinensis*, and pods from *Caesalpinia spinosa*. The European chestnut tree (*Castanea sativa*) and the oak species especially *Quercus montana* also produce hydrolyzable tannins in sufficient amount which are used in leather manufacture. Ellagitannins, a class of hydrolysable tannins produce ellagic acid on hydrolysis under acidic or alkaline conditions. Mostly these are present in angiosperms extractives. Ellagic acid and their derivatives have extensive applications as antioxidants, chelators, technical and biomedical applications. The possible applications viz. antibacterial, antifungal, antiviral, anti-inflammatory, hepato- and cardioprotective, chemopreventive, neuroprotective, anti-diabetic, gastroprotective, antihyperlipidemic, and antidepressant-like activities, among others have gained interest to researchers and reviewed for commercial exploitation [108, 109].

Condensed tannins (nonhydrolyzable tannin or proanthocyanidine) the larger polyphenol groups with high molecular weight upto 30000D compounds, form insoluble precipitates in aqueous solution and are the polymerization products of monomeric flavan–3-ol or flavan-3,4-diol precursors (**Figure 2c** and **d**) [110] which are joined through stable C-C bonds between C-4 and C-8 and between C-4 and C-6. Tannins are dynamically used in tanning of animal hides in the leather processing industry since 1960s, the beginning of the industry due to interaction and precipitation of the proteins [29], adhesive making (especially wood adhesives), fisheries, beverages manufacturing, animal feed, biosourced foams, wood preservatives, corrosion inhibitors, polyurethane surface coatings, epoxy adhesives, binders for Teflon coatings, as mineral absorption and protein, as iron gall ink production, adhesive production in wood-based industry, anti-corrosive chemical production, uranium recovering from seawater, and removal of mercury and methylmercury from solution. In continuum, tannins are also used as bioactive molecules in nutrition science, 3D printing and biomedical devices [109, 111]. Their presence in natural vegetable material has prompted scientific community for their industrial applications in many different ways. Since historical times their traditional use has allowed their further use after diverse chemical modification for various end use functionalized properties. The main inherent feature of the tannins is due to the presence of phenolic structure similar to synthetic phenols. Mostly the condensed tannins are polymers composed of falvan-3-ol monomers and are mainly extracted from bark and wood for commericial purposes. Structural diversity and functions of varied range of tannins are very well described elsewhere in the literature [112–116]. Tannins are extracted from plant material by simple methods. Nevertheless, there are various extraction processes were developed to isolate the tannins for diverse applications. However, the extraction process remains a challenge due to their heterogeneity character and compositions. Recently, various extraction processes, technological applications and their pros and cons were reviewed and appeared in leading scientific reports [117–119]. Due to similar tannin structural properties as that of synthetic phenols, the basis of wood adhesives was started in the middle of 1940s. The world first commercial wood adhesive credit goes to Australia in the 1960s using *Acacia mearnsii* (black wattle) bark tannins. The successes stories of producing tannin wood adhesive are continued till date with the advancements. The other species used in producing commercial taanins are *Pinus radiate* (radiate pine) bark [120].

Condensed tannins have been in industrial use for nearly 60 years as replacement for phenolic resins for wood based panels with high resistance against moisture and water as well as for boards with very low subsequent formaldehyde emission. Mimosa tannins, obtained from mimosa bark are usually well appreciated for its functional properties for wood adhesives. A wider industrial usage of tannins suffers from the limited availability of raw materials and high transportation costs. Only South Africa is the only actual producer of mimosa tannins on industrial scale [121]. Tannins react with formaldehyde the main crosslinker, and form hardened and crosslinked structures, similar to synthetic phenolic resins. The methylene bridges are formed between two tannin molecules. These methylene bridges are resistant to environmental factors against hydrolysis due to the strong stability of C-C bonds. Tannin-phenolic resin (tannin wt 40%) and sisal fibres (50 wt%) thermoset adhesives were successfully prepared and met all the required standards viz. Izod impact strength increased significatly. Further, it was also observed that sisal fibres and the tannin–phenolic thermosets have close values of the dispersive component and compatible interaction between the sisal fibres and the tannin–phenolic matrix at the interface [122]. A new source of tannin was also reported as a by product during catechin extraction process from a plant leaves (*Uncaria gambier*) extract and the tannin-phenol-formaldehyde wood adhesive was successfully prepared which met all the international standards [110]. Further, a new source of fibres, obtained from *Pinus roxburghii* needles, was also utilized in preparation of composites using phenolic resin adhesives [101]. Since, the early 1990s there are several scientific reports and reviews pertaining to natural fibres and tannin extracts from a natural sources viz. plants, agriculture wastes are available in the literature [122–131].

5.2 Biocomposites from fibres using enzyme and lignin

The study of enzymatic systems to activate the cellulosic fibres is a green alternative approach to other modifications for preparation of biocomposites and very well scientifically studied in last four decades to improve the surface, chemical, morphological and thermal properties of natural fibres as reinforcement materials. The enzymes offer an inexpensive and ecofriendly attractive option to improve the surfaces of natural fibres for composites.

Laccases (EC1.10.3.1) are multinuclear copper oxidases often called 'blue' oxidases that catalyze the oxidation of a wide range of susbstrates including phenols. Fungal laccases (benzenediol:oxygen oxidoreductase, EC1.10.3.2) are obtained from extractives of various fungal strains as an extracellular product. This enzyme is produced extensively in higher plants and fungi. The enzyme is produced by different genera of ascomycetes [132–134]; deuteromycetes [135, 136] and mainly from basidiomycetes [137]. The production and purification of biotechnological enzymes have been reviewed extensively [138–145] due to its overwhelming response. The first laccase was obtained from a Japanese lacquer tree (Rhus vernicifera), since then new fungal laccases from *Trametes versicolor*, *Polyporus pinsitus*, *Rhizocotonia solani* and from Ascomycetes *Myceliophthora* thermophila and *Sccytalidium thermophilum* etc. were obtained and studied extensively [142, 143, 145–159].

Commercially, laccases have been used for delignification of wood, production of ethyl alcohol and identification of morphine and codeine etc. among the various applications [142, 160–170]. Various delignification processes using fungi have been developed by the scientific community successfully. These enzymes were considered to be capable of Cα-Cβ cleavage of the side chain of lignin models and it was suggested that the enzymes participate in lignin degradation [171, 172]. The white rot fungi especially basidiomycetes degrade lignin in natural system more robustly than other organisms. They completely degrade lignin to carbon dioxide and water. The lignin degradation by white rot fungi was extensively studied earlier on *Phanerochaete chrysoporium* and *Sporotrichum pulverulentum* [173, 174]. The enzymes lignin peroxidase and manganese peroxidase were the first isolated from the *Phanerochaete chrysoporium* fungus culture. Lignin peroxidase is capable of ionizing non-phenolic aromatic substrates as an oxidizing peroxidase and produce aryl cation radicals [175, 176], whereas, manganese peroxidase does not degrade the non-phenolic parts of lignin in wood [177]. Further, it was observed that laccases exhibit strong catalytic activity and biotechnological applications in the bleaching of kraft pulp by depolymerising and solubilising lignin in the presence of mediator compounds [168, 178–182]. These mediator compounds were called laccase mediator systems (LMS). A number of possible mediator compounds have been searched and described for enhancing the activity of enzymes but mostly the ABTS (2,2'-azino-bis(3-ethylbenzothiazoline-6-sulfonic acid) and HBT (1-hydroxybenzotriazole) (**Figure 2e** and **f**), a derivative of benzotriazole have been used as the mediator systems [173, 183]. The oxidation of benzyl alcohols with $ABTS^{2+}$ (**Figures 3** and **4**) have been experimentally proved by [182–184]. The reaction mechanism of lignin degradation with LMS is complex due to the complex nature of LMS and various reactive centres on the lignin molecule. The simple reaction mechanism was postulated by Freudenreich [185] on the non-phenolic lignin model compound, veratryl alcohol and suggested a possible mechanism of delignification of residual lignin. The scientific kinetic studies of veratryl alcohol and benzyl alcohols have also been studied in laccase mediator systems [178, 184, 186, 187].

Laccases have been found to possess catalytic ability not only to degrade lignin and in delignification process for applications in biobleaching process but also observed as their involvement in the *in vivo* lignin biosynthesis and possibly in lignification woody tissues in higher plants [188–196]. The approach of oxidation and polymerisation of lignin by the enzymes [161, 162, 197] was advanced to another biotechnological application for compounded materials using wood fibres, lignin and enzyme laccase. The high tensile strength of the woody system or biocomposites is produced by the cellulose fibres and the pressure strength is produced by the lignin matrix which is a cementing material polymerised *in situ* and held cellulose fibres together resulting in high strength to the natural composites. Cellulose and hemicelluloses (**Figure 5a**) are macromolecules composed of sugar molecules. Cellulose is composed of only glucose molecules having β-1-4 linkage (**Figure 5b**), whereas hemicelluloses composed of different sugar monomers aligned in a definite fashion. Cellulose formation by a single glucose molecule in plant cell requires four enzymes and biosynthesis of lignin composed of phenol units utilizes peroxidases and phenoloxidases (Laccases) [198–201]. The monomeric units of lignin comprised of coumaryl alcohol (H-lignin) present in grasses and agriculture crops, coniferyl alcohol (G-lignin) present in all species, dominant in conifers and syringyl alcohol (S-lignin) present in hard wood species up to 40% (**Figure 6**).

The production of composites emanates the same basic principle as the formation of natural wood: wood is processed and fragmented into fibres and small pieces as per need of the required composites. Fibres, isolated from soft or hardwood, fibre bearing species or agriculture wastes, are embedded into a matrix. The matrix or binder may be a urea-formaldehyde, phenol-formaldehyde, resorcinol-formaldehyde, isocynates or in a combination as per requirement of the composites.

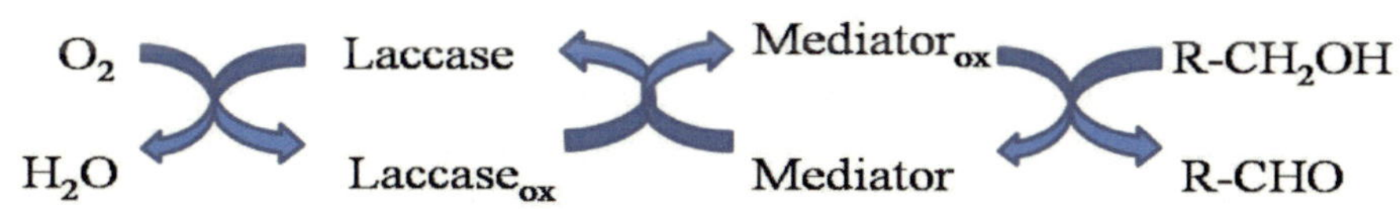

Figure 3.
Laccase-catalyzed oxidation of veratryl alcohol in the presence of a mediator.

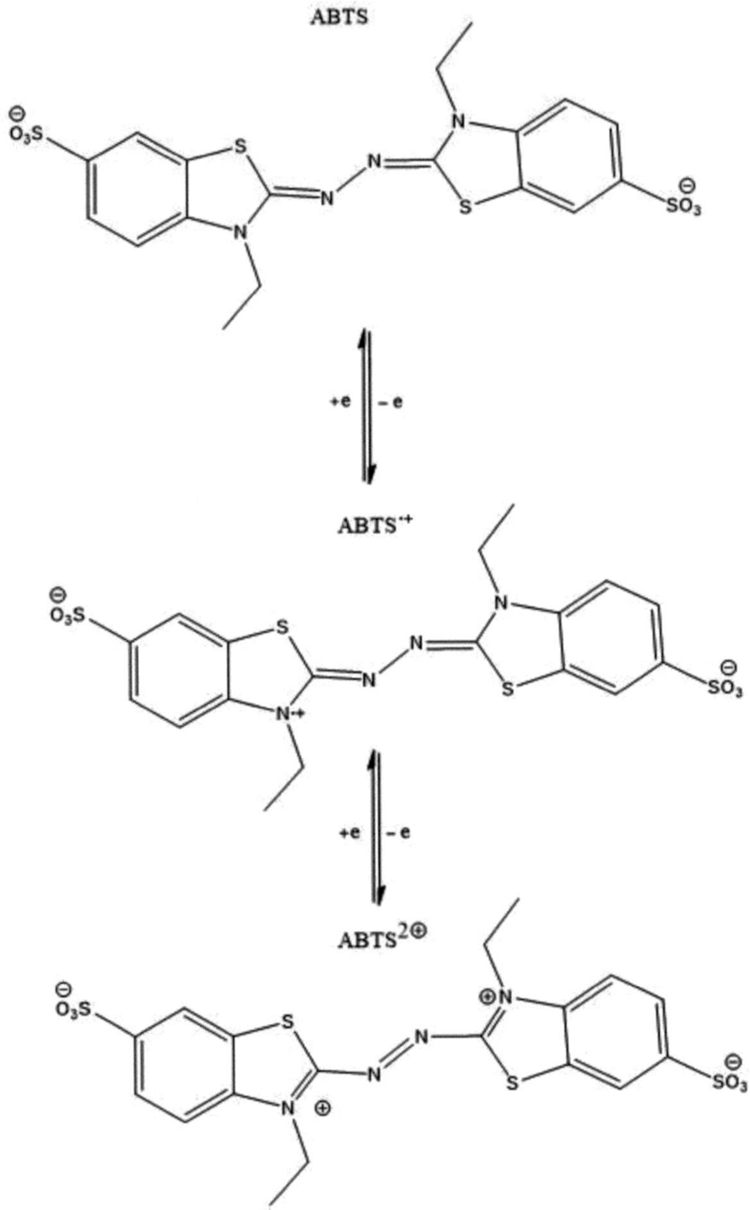

Figure 4.
Oxidation of ABTS by Laccase enzyme (blue colour, λ_{max}753nm).

The postulated theories of delignification of lignin, lignifications of woody tissues, activation of the surface of fibres possessing lignin, by the peroxidases enzymes have been successfully applied to prepare green biocomposites. The use of enzyme for bonding in the wood was first suggested by Nimz [202]. In continuum, several scientific communities have been engaged in producing biocomposites using laccase peroxidase enzyme. Wood fibres are incubated for a certain time with phenoloxidas laccase enzyme and lignin crust on the fibre surface gets activated and oxidized. Activated fibres are compressed by operating standard operating conditions of pressure, temperature etc, and binderless fibre boards may be prepared as per standards [170, 200, 201, 203–207]. The utilization of peroxidases in production of biocomposites was also applied to fibres in last two decades [208]. Cellulose fiber enzyme

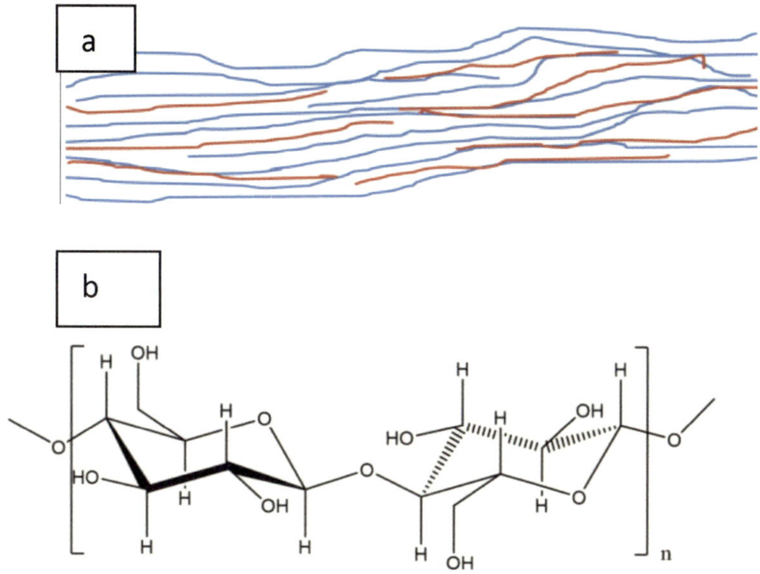

Figure 5.
(a) Schematic diagram of cellulose and hemicelluloses in cellulose microfibrils arrangement, blue lines: cellulose; red line: hemicelluloses (b) Chemical Structure of cellulose.

composites [207], hemp fibre reinforced composites using enzyme and chelators [209], polypropylene composites using abaca fibre [210], sisal fibre/phenolic resin composites [211], laccase-treated kenaf fibre reinforced composites with polypropylene and maleic anhydride grafted polypropylene as coupling agents [212], rubber wood fibreboards [205], laccase-mediated grafting dodecyl gallate (DG) on the jute fiber composites [213], banana/polypropylene composites [214, 215], coconut fibre composites [216], natural fiber medium density fibreboard [217], jute polypropylene composites [218, 219], flax fibre epoxy Composites [220, 221] were successfully prepared and studied for increasing mechanical properties and interfacial adhesion of the biocomposites. These all studies indicated that enzymes have the potential ability to modify the surface properties of fibres as being utilized in production of biocomposites. The formation of biocomposites has been shown in graphical representation (**Figure 7**) [163, 165, 200, 201, 203, 204, 222–225].

6. Opportunities and future perspectives

In this chapter, we have underlined and discussed the different sources of natural fibres, their properties and the effect of treatments on natural fibres, etc. and further their effective use as reinforcement for polymer composite materials. Natural fibres are lucrative and worthwhile option for biocomposites. However, limitations such as poor thermal stability, moisture absorption and poor compatibility with polymeric matrices are challenges that need to be resolved.

There are a large number of fibres obtained from the natural resources; intriguingly only few of these fibres have been studied in detail for reinforcement of bio-composite materials and other industrial and traditional applications. In the present chapter popular natural fibres have been discussed as reinforced composites materials with combination of synthetic and natural polymers as modified matrix. Among the most popular natural fibres; flax, jute, hemp, sisal, ramie, and kenaf fibres were extensively studied and employed in different applications as reinforced

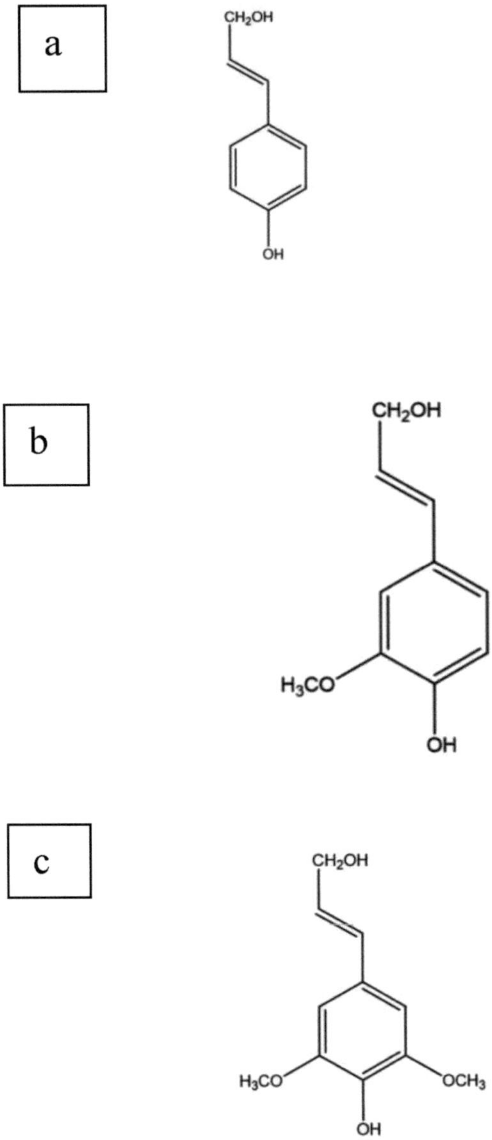

Figure 6.
Monomeric units of Lignin (a) Coumaryl alcohol (H-lignin) (b) Coniferyl alcohol (G-lignin) (c) Syringyl alcohol (S-lignin).

materials. But due to environmental and economic concern other fibres from natural resources such as pine, bagasse, pineapple leaf, coir, oil palm, banana, and agriculture residues are acquiring interest for various value added applications due to their inherent and diverse physical properties. Merits and demerits of the natural fibres and their inherent properties mainly influence the mechanical properties of bio-composites due to interfacial adhesion between the fibre and synthetic matrix.

Variability in natural fibres such as processing conditions, fibre diameter and length, lumen diameter, presence of other compounds such as amount of lignin and hemicelluloses needs to be standardising for the processing of particular fibres. High qualities of fibres are required to increase the potential of fibres as reinforcement materials. Maleated coupling agents are extensively used to enhance the composites strength using fibre as reinforcement material. These agents are used as

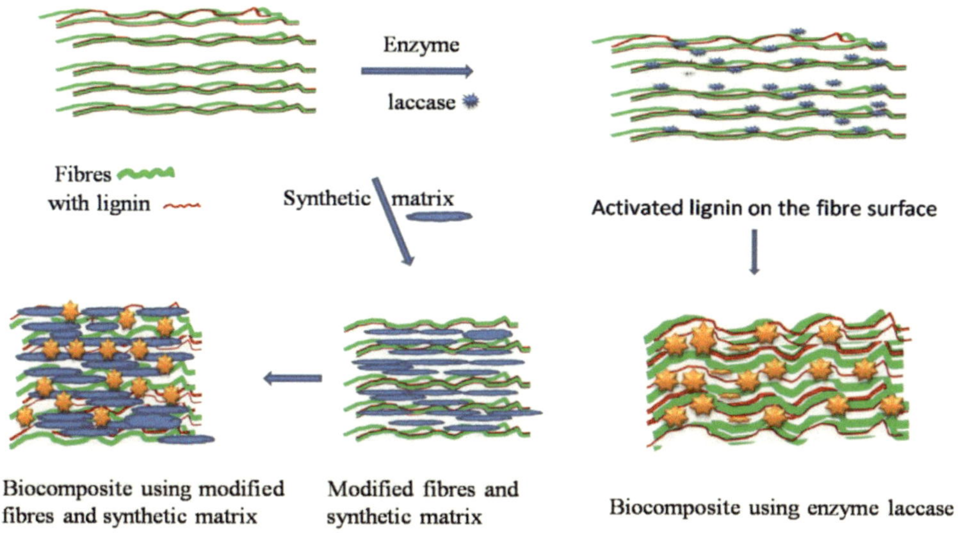

Figure 7.
Graphical representation of Biocomposites.

couplers and bind synthetic matrix and functional surface of fibres and economical in processing. Further, using couplers the strength of fibres increased which lead to increase interfacial adhesion between two dissimilar components. However, maleated couplers illustrate superior performances with polypropylenes, polylactic acid and other polyolefins etc. Further, scientific inputs are required to improve the strength of biocomposites using maleated couplings by incorporating varied fibres.

The diverse ranges of fibres are required to investigate the quantification of residual lignin on the fiber surface and optimization of fiber isolation parameters since during the processing of fibres the amount of residual lignin may be different to the fibres isolated from the same resource. Further, the constituents and structure of lignin is also different in the fibres sourced material viz. soft and hard woods and vegetable crops and agriculture residues. Therefore, proper attention is required to investigate the activation of fibre surface and binder in the biocomposites similar to natural composites. In continuum, diverse range of appropriate laccase mediator systems (LMS) needs attention for biocomposites.

Green composites may be a suitable alternative for petroleum-based synthetic non-environment friendly materials by using enzymes especially 'laccases' one of the most ancient and efficient enzymes with promising future applications. The high reduction potential of laccases has led the vast industrial applicability, despite this, laccase potentialities are not fully exploited due to large-scale production, cost and efficiency. Systemetic progress has been made over the last three decades to enhance the utilization of laccase enzyme for various biotechnological applications and it is expected that laccases will be able to compete with other processes of bio-composites. Thus, scientific efforts are need of the hour in order to achieve the economical production of the biocatalyst, development of optimum production conditions like pH, temperature, medium composition and efficient mediator systems and further utilization in lignin activation of fibres.

In view of the above discussion, the following activities may be expedient in bio-composites development from natural resources.

- Identification and search of new fibres with better inherent properties and compatibility to the synthetic matrices.

- Agriculture residues consisting of sufficient amount of celluloses.
- Utilization of tannins and non toxic small aldehydes for making wood adhesives.
- Search of coupling agents and other bonding agents.
- Efficient production of laccases and other biotechnological economical enzymes from various fungal resources.
- Utilization of waste lignin from the paper industry and activation thereof using enzymes for lignin celluloses complexes.
- Activation of lignin on the fibre surface to act as binder and enzyme mediator systems to form efficient lignin cellulose complexes.
- Utilization of long and short fibres simultaneously to avoid voids for increasing the compactness and stiffness of bio-composites.
- Lignin starch/celluloses complexes for biodegradable plastics.
- Fire and water proof green bio-composites using natural fibres for interior design of automotive, aerospace and applications in construction industry.
- Utilization of laccases, celluloses, starches and lignin for various food grade composites as an alternative to plastic.

Acknowledgements

The author is thankful to my parent organization Forest Research Institute, Indian Council of Forestry Research and Education, Dehradun India, where I cultured myself in the Chemistry of Forest Products in the amalgamation of diverse disciplines under one umbrella. My heartfelt thanks are also due to Prof. (L) Aloys Huttermann, Ex-Direktor, Forstbotanisch Institute Der Universitat, Gottingen, Germany for earning the concepts of biocomposites using Enzymes and Prof. Michael J Kennedy Ex-Leader, Forestry Science, Department of Agriculture and Fisheries (Formerly: Department of Primary Industries and Fisheries), Brisbane, Queensland, Australia, for utilization of forest biomass wastes.

Conflicts of interest

The author declares no conflict of interest.

Author details

Pradeep Sharma
Chemistry and Bioprospecting Division, Forest Research Institute, Indian Council of Forestry Research and Education, Dehradun, India

*Address all correspondence to: drsharma27@gmail.com

IntechOpen

© 2021 The Author(s). Licensee IntechOpen. This chapter is distributed under the terms of the Creative Commons Attribution License (http://creativecommons.org/licenses/by/3.0), which permits unrestricted use, distribution, and reproduction in any medium, provided the original work is properly cited.

References

[1] Lovrić, M., Da Re, R., Vidale, E., Prokofieva, I., Wong, J., Pettenella, D., ... & Mavsar, R. (2020). Non-wood forest products in Europe–A quantitative overview. Forest Policy and Economics, 116, 102175.

[2] Panayotou, T., & Ashton, P. (1992). Not by timber alone: economics and ecology for sustaining tropical forests. Ch 5, 70-109, Island Press, Washington DC.

[3] Talukdar, N. R., Choudhury, P., Barbhuiya, R. A., & Singh, B. (2021). Importance of Non-Timber Forest Products (NTFPs) in rural livelihood: A study in Patharia Hills Reserve Forest, northeast India. Trees, Forests and People, 3, 100042. https://doi.org/10.1016/j.tfp.2020.100042.

[4] Chopra, K. (1993) The value of NTFP: an estimation for tropical deciduous forest in India Econ. Bot., 47, 251-257

[5] Vantomme, P., Markkula, A., & Leslie, R. N. (2002). Non-wood forest products in 15 countries of tropical Asia: a regional and national overview. FAO Regional Office for Asia and the Pacific, ISBN : 974906660X.

[6] Weiss, G., Emery, M. R., Corradini, G., & Živojinović, I. (2020). New values of non-wood forest products. Forests, 11(2), 165.

[7] Keča, L. J., Keča, N., & Rekola, M. (2013). Value chains of Serbian non-wood forest products. International Forestry Review, 15(3), 315-335.

[8] Kurttila, M., Pukkala, T., & Miina, J. (2018). Synergies and trade-offs in the production of NWFPs predicted in boreal forests. Forests, 9(7), 417.

[9] Rajendra, K. C. (2018). Contribution of NWFPs in National Economy. *Banko Janakari*, *28*(2), 1-2.

[10] Chandramohan, D., & Marimuthu, K. (2011). A review on natural fibres. International Journal of Research and Reviews in Applied Sciences, 8(2), 194-206.

[11] Sanjay, M. R., Arpitha, G. R., Naik, L. L., Gopalakrishna, K., & Yogesha, B. (2016). Applications of natural fibres and its composites: An overview. Natural Resources, 7(3), 108-114.

[12] Lobovikov, M., (2003). Bamboo and rattan products and trade. J. Bamboo Ratt. 2 (4), 397-406.

[13] Anonymous, (2019). Indian State of Forest Report, Bamboo Resources of the Country. Forest Survey of India 16th ed. Chapter 8, Dehradun, Uttarakhand, India

[14] Bystriakova, N., (2003). Bamboo Biodiversity: Information for Planning Conservation and Management in the Asia-pacific Region. UNEP World Conservation Monitoring Centre.

[15] Bystriakova, N., Kapos, V., Lysenko, I., (2004). Bamboo biodiversity: Africa, Madagascar and the Americas. UNEP/ Earthprint.

[16] Sawarkar, A. D., Shrimankar, D. D., Kumar, A., Kumar, A., Singh, E., Singh, L. & Kumar, R. (2020). Commercial clustering of sustainable bamboo species in India. Industrial Crops and Products, 154, 112693. https://doi.org/10.1016/j.indcrop.2020.112693

[17] Seethalakshmi, K. K., Kumar, M. M., Pillai, K. S., & Sarojam, N. (1998). Bamboos of India: A compendium (Vol. 17). Brill.

[18] Sharma, Y. M. L. (1980). Bamboos in the Asia Pacific Region. In Bamboo research in Asia: proceedings of a workshop held in Singapore, 28-30 May 1980. IDRC, Ottawa, ON, CA.

[19] Borah, E. D., Pathak, K. C., Deka, B., Neog, D., & Borah, K. (2006). Utilization aspects of Bamboo and its market value. The Indian Forester, 423-427.

[20] Farrelly, D. (1996). The Book of Bamboo: A Comprehensive Guide to This Remarkable Plant, Its Uses, and Its History. Thames and Hudson Ltd.

[21] Hunter, I.R. (2003) Bamboo resources, uses and trade: The future? J. Bamboo Ratt. 2, 319-326.

[22] Liese, W., Köhl, M. (2015) Bamboo: the Plant and Its Uses, 299-346, Springer.

[23] Amada S, Ichikawa Y, Munekata T, Nagase Y, Shimizu H. (1997) Fiber texture and mechanical graded structure of bamboo. Composites Part B 1997; 28B:13-20.

[24] Amada S, Munekata T, Nagase Y, Ichikawa Y, Kirigal A, Yang Z. (1996) The mechanical structures of bamboos in view of functionally gradient and composite materials. J Comput Mater 30(7):800-19.

[25] Chuma S, Hirohashi M, Ohgama T, Kasahara Y. (1990) Composite structure and tensile properties of mousou bamboo. J Mater Soc Japan, 39:847-51.

[26] Lakkard S.C, Patel J.M. (1981) Mechanical properties of bamboo, a natural composite. J Fibre Sci Tec 14:319-22.

[27] Nogata F. & Takahashi H. (1995) Intelligent functionally graded material: bamboo. Compos Eng 5(7):743-51.

[28] Rao, K. M. M., & Rao, K. M. (2007). Extraction and tensile properties of natural fibres: Vakka, date and bamboo. *Composite structures*, 77(3), 288-295.

[29] Haslam, E. "In Biochemistry of Plants: A comprehensive Treatise, Vol 7, Secondary Plant Products, Ed. PK Stumpf and EE Conn." (1981): 527-556.

[30] Haslam, E. (1979). Vegetable tannins. In Biochemistry of plant phenolics (pp. 475-523). Springer, Boston, MA.

[31] Gupta, V. K. (2019). Application of Natural Dyes, ch6, In Fundamentals of natural dyes and its application on textile substrates. Chemistry and technology of natural and synthetic dyes and pigments. DOI: 10.5772/intechopen.89964.

[32] Saxena, S., & Raja, A. S. M. (2014). Natural dyes: sources, chemistry, application and sustainability issues. In Roadmap to sustainable textiles and clothing (pp. 37-80). Springer, Singapore.

[33] Gholampour, A., & Ozbakkaloglu, T. (2020). A review of natural fiber composites: Properties, modification and processing techniques, characterization, applications. Journal of Materials Science, 55(3), 829-892. https://doi.org/10.1007/s10853-019-03990-y

[34] Rajisha, K. R., Deepa, B., Pothan, L. A., & Thomas, S. (2011). Thermomechanical and spectroscopic characterization of natural fibre composites. *Interface Engineering of Natural Fibre Composites for Maximum Performance*, 241-274. https://doi.org/10.1533/9780857092281.2.241

[35] Ticoalu, A., Aravinthan, T., & Cardona, F. (2010). A review of current development in natural fiber composites for structural and infrastructure applications. In *Proceedings of the southern region engineering conference (SREC 2010)* (pp. 113-117). Engineers Australia.

[36] Bachtiar D., Sapuan S. M., Zainudin E. S., Khalina A., and Dahlan K. Z. M. (2009) "The tensile

properties of single sugar palm (Arenga pinnata) fibre," presented at the 9th National Symposium on Polymeric Materials (NSPM 2009).

[37] Beckermann G. W. and. Pickering, K. L. (2008) "Engineering and evaluation of hemp fibre reinforced polypropylene composites: Fibre treatment and matrix modification," Composites: Part A, vol. 39, pp. 979-988.

[38] Brouwer, W. "Natural fiber composites in structural components: Alternative applications for sisal? Common Fund for Commodities - Alternative Applications for Sisal and Henequen," Technical Paper No. 14, Proceedings of a Seminar held by the Food and Agriculture Organization of the UN (FAO) and the Common Fund for Commodities (CFC), 2000.

[39] Jose, S., Rajna, S., & Ghosh, P. (2017). Ramie fibre processing and value addition. Asian Journal of Textile, 7(1), 1-9.

[40] Jose, S., Salim, R., & Ammayappan, L. (2016). An overview on production, properties, and value addition of pineapple leaf fibers (PALF). Journal of Natural Fibers, 13(3), 362-373.

[41] Ketnawa, S., P. Chaiwut, and S. Rawdkuen. (2012). Pineapple wastes: A potential source for bromelain extraction. Food and Bio-Products Processing 90: 385-391. doi:10.1016/j.fbp.2011.12.006

[42] Mohanty K., Misra, M. and Hinrichsen, G. (2000) Biofibres, biodegradable polymers and biocomposites: An overview, Macromolecular Materials and Engineering, vol. 276-277, pp. 1-24.

[43] Mwaikambo, L. Y. (2006). Review of the history, properties and application of Plant fibres. African Journal of Science and Technology, Science and Engineering Series 7: 120-133.

[44] Siakeng, R.; Jawaid, M.; Ari_n, H.; Sapuan, S.M.; Asim, M.; Saba, N. (2018) Natural fiber reinforced polylactic acid composites: A review. Polym. Compos., 40, 446-463.

[45] Sreenivas. H.T., Krishnamurthy, N., & Arpitha, G. R, A. (2020). A comprehensive review on light weight kenaf fiber for automobiles. International Journal of Lightweight Materials and Manufacture, 3(4), 328-337.

[46] Ayadi, R., Hanana, M., Mzid, R., Hamrouni, L., Khouja, M. L., & Salhi Hanachi, A. (2017). Hibiscus cannabinus L.–kenaf: a review paper. Journal of Natural Fibers, 14(4), 466-484.

[47] Bekele, L. D., Zhang, W., Liu, Y., Duns, G. J., Yu, C., Jin, L., ... & Chen, J. (2017). Preparation and characterization of lemongrass fiber (Cymbopogon species) for reinforcing application in thermoplastic composites. BioResources, 12(3), 5664-5681.

[48] Draman, S. F. S., Daik, R., Latif, F. A., & El-Sheikh, S. M. (2014). Characterization and thermal decomposition kinetics of kapok (Ceiba pentandra L.)–based cellulose. BioResources, 9(1), 8-23.

[49] Khalil, H. A., Yusra, A. I., Bhat, A. H., & Jawaid, M. (2010). Cell wall ultrastructure, anatomy, lignin distribution, and chemical composition of Malaysian cultivated kenaf fiber. Industrial Crops and Products, 31(1), 113-121.

[50] Koohestani, B., Darban, A.K., Mokhtari, P. (2019) Comparison of different natural fiber treatments: a literature review. Int. J. Environ. Sci. Technol. 16, 629-642. https://doi.org/10.1007/s13762-018-1890-9

[51] Faruk, O., Bledzki, A. K., Fink, H. P., & Sain, M. (2014). Progress report

on natural fiber reinforced composites. Macromolecular Materials and Engineering, 299(1), 9-26. https://doi.org/10.1002/mame.201300008

[52] Jawaid, M. H. P. S., & Khalil, H. A. (2011). Cellulosic/synthetic fibre reinforced polymer hybrid composites: A review. Carbohydrate polymers, 86(1), 1-18.. https://doi.org/10.1016/j.carbpol.2011.04.043

[53] Ramamoorthy, S. K., Skrifvars, M., & Persson, A. (2015). A review of natural fibres used in biocomposites: plant, animal and regenerated cellulose fibres. *Polymer Reviews*, *55*(1), 107-162. https://doi.org/10.1080/15583724.2014.971124

[54] John, M. J., & Anandjiwala, R. D. (2008). Recent developments in chemical modification and characterization of natural fiber-reinforced composites. Polymer composites, 29(2), 187-207. https://doi.org/10.1002/pc.20461

[55] Shalwan and Yousif, B. F. (2013) "In state of art: mechanical and tribological behaviour of polymeric composites based on natural fibres," Materials & Design, vol. 48, pp. 14-24.

[56] Sreekala M. S. and Thomas S. (2003) "Effect of fibre surface modification on water-sorption characteristics of oil palm fibres," *Composites Science and Technology*, vol. 63, no. 6, pp. 861-869.

[57] Cruz, J., & Fangueiro, R. (2016). Surface modification of natural fibers: a review. Procedia Engineering, 155, 285-288.

[58] Bledzki, A. K., & Gassan, J. (1999). Composites reinforced with cellulose based fibres. Progress in polymer science, 24(2), 221-274. https://doi.org/10.1016/S0079-6700(98)00018-5

[59] Costa, C. G., Bom, L. F. R. P., Martins, C. R., da Silva, C. F., & de Moraes, M. A. (2020). (Bio) composites of chitin/chitosan with natural fibres. In Handbook of Chitin and Chitosan (pp. 273-298). Elsevier. 10.1016/B978-0-12-817968-0.00009-3

[60] Semsarzadeh M. A., (1986) Fibre matrix interaction in jute-reinforced polyester resin, Polym. Comp., 7(1), 23-25. https://doi.org/10.1080/03602559108019223

[61] Haig Zeronian, S., Kawabata, H., & Alger, K. W. (1990). Factors affecting the tensile properties of nonmercerized and mercerized cotton fibres. Textile Research Journal, 60(3), 179-183.

[62] Wulfhorst B, Tetzlaff G, Kaldenhoff R. (1992) Techn Text; 35(3):10-11.

[63] Belgacem, M. N., Bataille, P., & Sapieha, S. (1994). Effect of corona modification on the mechanical properties of polypropylene/cellulose composites. Journal of applied polymer science, 53(4), 379-385. https://doi.org/10.1002/app.1994.070530401

[64] Dong, S., Sapieha, S., & Schreiber, H. P. (1992). Rheological properties of corona modified cellulose/polyethylene composites. Polymer Engineering & Science, 32(22), 1734-1739. https://doi.org/10.1002/pen.760322212

[65] Sakata, I., Morita, M., Tsuruta, N., & Morita, K. (1993). Activation of wood surface by corona treatment to improve adhesive bonding. *Journal of Applied Polymer* Science, 49(7), 1251-1258.

[66] Gao, S., & Zeng, Y. (1993). Surface modification of ultrahigh molecular weight polyethylene fibres by plasma treatment. I. Improving surface adhesion. Journal of applied polymer science, 47(11), 2065-2071. https://doi.org/10.1002/app.1993.070471116

[67] Ragoubi, M., Bienaimé, D., Molina, S., George, B., & Merlin, A. (2010).

Impact of corona treated hemp fibres onto mechanical properties of polypropylene composites made thereof. *Industrial Crops and Products*, *31*(2), 344-349. https://doi.org/10.1016/j.indcrop.2009.12.004

[68] Ragoubi, M., George, B., Molina, S., Bienaimé, D., Merlin, A., Hiver, J. M., & Dahoun, A. (2012). Effect of corona discharge treatment on mechanical and thermal properties of composites based on miscanthus fibres and polylactic acid or polypropylene matrix. *Composites Part A: Applied Science and Manufacturing*, *43*(4), 675-685. https://doi.org/10.1016/j.compositesa.2011.12.025

[69] Oudrhiri F. H., Merbahi, N., Oushabi, A., Elfadili, M.H., Kammouni, A., Oueldna, N., Effects of corona discharge treatment on surface and mechanical properties of Aloe vera fibres, Materials Today: Proceedings, Volume 24, Part 1, (2020), Pages 46-51, ISSN 2214-7853, https://doi.org/10.1016/j.matpr.2019.07.527.

[70] Danmei Sun (2016), Surface Modification of Natural Fibres Using Plasma Treatment In book: Biodegradable Green Composites, Chapter 2: Surface modification of natural fibres using plasma treatment, Publisher: John Wiley & Sons, Editors: Kalia S. DOI: 10.1002/9781118911068.

[71] Sobczyk-Guzendaa, A; Szymanowskia, H.; Jakubowskia, W., Błasińskab, A.; Kowalskia J. and Gazicki-Lipmana, M., (2013). Morphology, photocleaning and water wetting properties of cotton fabrics, modified with titanium dioxide coatings synthesized with plasma enhanced chemical vapor deposition technique, Surface and Coatings Technology, 217(25), pp. 51-57.

[72] Sun, D. and Stylios, G., (2005). Investigating the plasma modification of natural fibre fabrics – the effect on fabric surface and mechanical properties, Textile Research Journal, 75 (9), pp. 639-645.

[73] Mukhopadhyay, S.; Pal, R.; Narula, V.; Mayank, M., (2013). A study of interface behavior in sisal fibre composites - Single fibre pull out test. Indian Journal of Fibre & Textile Research, 38 (1), pp. 87-91.

[74] Bozaci E.; Sever K.; Sarikanat M.; Seki Y.; Demir A.; Ozdogan E. and Tavman, I., (2013). Effects of the atmospheric plasma treatments on surface and mechanical properties of flax fiber and adhesion between fiber–matrix for composite materials, Composites: Part B, 45, pp. 565-572.

[75] Sarikanat, M., Seki, Y., Sever, K., Bozaci, E., Demir, A., & Ozdogan, E. (2016). The effect of argon and air plasma treatment of flax fiber on mechanical properties of reinforced polyester composite. *Journal of Industrial Textiles*, 45(6), 1252-1267.

[76] Li, X., Tabil, L. G., & Panigrahi, S. (2007). Chemical treatments of natural fiber for use in natural fiber-reinforced composites: a review. Journal of Polymers and the Environment, 15(1), 25-33. 10.1007/s10924-006-0042-3

[77] Zhou, M., Yan, J., Li, Y., Geng, C., He, C., Wang, K., & Fu, Q. (2013). Interfacial strength and mechanical properties of biocomposites based on ramie fibers and poly (butylene succinate). RSC advances, 3(48), 26418-26426.

[78] Demir, A., Seki, Y., Bozaci, E., Sarikanat, M., Erden, S., Sever, K., & Ozdogan, E. (2011). Effect of the atmospheric plasma treatment parameters on jute fabric: the effect on mechanical properties of jute fabric/polyester composites. Journal of Applied Polymer Science, 121(2), 634-638.

[79] Gibeop, N., Lee, D. W., Prasad, C. V., Toru, F., Kim, B. S., & Song, J. I.

(2013). Effect of plasma treatment on mechanical properties of jute fiber/poly (lactic acid) biodegradable composites. Advanced Composite Materials, 22(6), 389-399.

[80] Kafi, A. A., Magniez, K., & Fox, B. L. (2011). A surface-property relationship of atmospheric plasma treated jute composites. Composites science and technology, 71(15), 1692-1698. https://doi.org/10.1016/j.compscitech.2011.07.011

[81] Kim, B. S.; Nguyen, M. H.; Hwang, B. S. and Lee, S., (2008). Effect of plasma treatment on the mechanical properties of natural fibre/PP composites, in: De Wilde, W.P., Brebbia, C.A. (Eds.), High performance structures and materials IV. WIT transactions on built environment, WIT PRESS, Southampton UK, pp. 159-166.

[82] Sathish, S., Karthi, N., Prabhu, L., Gokulkumar, S., Balaji, D., Vigneshkumar, N., ... & Dinesh, V. P. (2021). A review of natural fiber composites: Extraction methods, chemical treatments and applications. *Materials Today: Proceedings*. https://doi.org/10.1016/j.matpr.2020.12.1105

[83] Goyal, P., Kumar, V. and Sharma, P. (2008) Cyanoethylation of Tamarind kernel Powder. Starch/Starke, 60 (1), 41-47.

[84] Goyal, P., Kumar, V. and Sharma, Pradeep (2007) Carboxymethylation of Tamarind kernel Powder. Carbohydrate Polymers, 69, 251-255.

[85] Gupta, S., Sharma, P. and Soni, P.L. (2004) Carboxymethylation of Cassia occccidentalis seed gum. J. Applied Polymer Science, 94(4) 1606-1611

[86] Gupta, S., Sharma, P. and Soni, P.L. (2005) Chemical modification of Cassia occidentalis seed gum: Carbamoylethylation. Carbohydrate Polymers, 59, 501-506.

[87] Kabir, M. M., Wang, H., Lau, K. T., & Cardona, F. (2012). Chemical treatments on plant-based natural fibre reinforced polymer composites: An overview. Composites Part B: Engineering, 43(7), 2883-2892. https://doi.org/10.1016/j.compositesb.2012.04.053

[88] Karaki, N., Aljawish, A., Humeau, C., Muniglia, L., & Jasniewski, J. (2016). Enzymatic modification of polysaccharides: Mechanisms, properties, and potential applications: A review. Enzyme and Microbial Technology, 90, 1-18.

[89] Keener, T. J., Stuart, R. K., & Brown, T. K. (2004). Maleated coupling agents for natural fibre composites. Composites part A: applied science and manufacturing, 35(3), 357-362. https://doi.org/10.1016/j.compositesa.2003.09.014

[90] Li, S., Xiong, Q., Lai, X., Li, X., Wan, M., Zhang, J., ... & Lin, Y. (2016). Molecular modification of polysaccharides and resulting bioactivities. Comprehensive Reviews in Food Science and Food Safety, 15(2), 237-250.

[91] Sharma D., Kumar V., Nautiyal R., Sharma P. (2020a) 'Synthesis and characterization of quaternized Cassia tora gum using Taguchi L'16 approach', Carbohydrate Polymers, Volume 232, Article115731. (https://doi.org/10.1016/j.carbpol.2019.115731.

[92] Sharma D., Kumar V., Sharma P. (2020b) Application, synthesis, and characterization of cationic galactomannan from ruderal species as wet strength additive and flocculating agent" ACS Omega DOI 10.1021/acsomega.0c03408.

[93] Sharma P., Gupta S., Soni P.L. and Kumar V. (2020) Ce(IV)-Ion Initiated grafting of 1,3 galactomannan with acrylonitrile. Journal of Macromolecular

Science. Part A. Pure and Applied Chemistry, Volume 57(&) 519-530. https://doi.org/10.1080/10601325.2020.1735254.

[94] Tyagi R., Sharma P., Nautiyal R., Lakhera A. K., and Kumar V. (2020). Synthesis of quaternised guar gum using Taguchi L (16) orthogonal array. Carbohydrate Polymers, 116136. https://doi.org/10.1016/j.carbpol.2020.116136 (Volume 237, Article, 116136, ISSN 0144-8617).

[95] Sahu, P., & Gupta, M. K. (2020). A review on the properties of natural fibres and its bio-composites: Effect of alkali treatment. Proceedings of the Institution of Mechanical Engineers, Part L: Journal of Materials: Design and Applications, 234(1), 198-217.

[96] Van de Weyenberg, I., Truong, T. C., Vangrimde, B., & Verpoest, I. (2006). Improving the properties of UD flax fibre reinforced composites by applying an alkaline fibre treatment. *Composites Part A: Applied Science and Manufacturing*, 37(9), 1368-1376. https://doi.org/10.1016/j.compositesa.2005.08.016

[97] Pereira, P. H. F., Rosa, M. D. F., Cioffi, M. O. H., Benini, K. C. C. D. C., Milanese, A. C., Voorwald, H. J. C., & Mulinari, D. R. (2015). Vegetal fibres in polymeric composites: a review. Polímeros, 25(1), 9-22. http://dx.doi.org/10.1590/0104-1428.1722

[98] Hossain, M. K., Karim, M. R., Chowdhury, M. R., Imam, M. A., Hosur, M., Jeelani, S., & Farag, R. (2014). Comparative mechanical and thermal study of chemically treated and untreated single sugarcane fiber bundle. Industrial Crops and Products, 58, 78-90. ttps://doi.org/10.1016/j.indcrop.2014.04.002

[99] Yanjun Xie, Callum A.S. Hill, Zefang Xiao, Holger Militz, Carsten Mai, (2010) Silane coupling agents used for natural fiber/polymer composites: A review, Composites Part A: Applied Science and Manufacturing, Volume 41, Issue 7,2010, pp 806-819,ISSN 1359-835X, https://doi.org/10.1016/j.compositesa.2010.03.005.

[100] Valadez-Gonzalez, A., Cervantes-Uc, J. M., Olayo, R., & Herrera-Franco, P. J. (1999). Chemical modification of henequen fibres with an organosilane coupling agent. *Composites Part B: Engineering*, 30(3), 321-331. https://doi.org/10.1016/S1359-8368(98)00055-9

[101] Sinha P., Mathur S., Sharma P., Kumar, V. (2016) Potential of Pine Needles for PLA based composites, Polymer Composites (DOI 10.1002/pc.24074).

[102] Aggarwal, P. K., Raghu, N., Karmarkar, A., & Chuahan, S. (2013). Jute–polypropylene composites using m-TMI-grafted-polypropylene as a coupling agent. Materials & Design, 43, 112-117.

[103] Han, G., Lei, Y., Wu, Q., Kojima, Y., & Suzuki, S. (2008). Bamboo–fiber filled high density polyethylene composites: effect of coupling treatment and nanoclay. Journal of Polymers and the Environment, 16(2), 123-130. https://doi.org/10.1007/s10924-008-0094-7

[104] Raghu, N., Kale, A., Chauhan, S., & Aggarwal, P. K. (2018). Rice husk reinforced polypropylene composites: Mechanical, morphological and thermal properties. *Journal of the Indian Academy of Wood Science*, 15(1), 96-104.

[105] Liu, W., Chen, T., & Qiu, R. (2014). Effect of fiber modification with 3-isopropenyl-dimethylbenzyl isocyanate (TMI) on the mechanical properties and water absorption of hemp-unsaturated polyester (UPE) composites. Holzforschung, 68(3), 265-271.

[106] Singh, H., Singh, J. I. P., Singh, S., Dhawan, V., & Tiwari, S. K. (2018). A brief review of jute fibre and its composites. Materials Today: Proceedings, 5(14), 28427-28437.

[107] Nandi, A., Kale, A., Raghu, N., Aggarwal, P. K., & Chauhan, S. S. (2013). Effect of concentration of coupling agent on mechanical properties of coir–polypropylene composite. Journal of the Indian Academy of Wood Science, 10(1), 62-67.

[108] Evtyugin, Dmitry D.; Magina, Sandra; Evtuguin, Dmitry V. (2020). "Recent Advances in the Production and Applications of Ellagic Acid and Its Derivatives. A Review" Molecules 25, no. 12: 2745. https://doi.org/10.3390/molecules25122745

[109] Lescano, C. H., de Lima, F. F., Caires, A. R. L., & de Oliveira, I. P. (2019). Polyphenols Present in Campomanesia Genus: Pharmacological and Nutraceutical Approach. In Polyphenols in Plants (pp. 407-420). Academic Press.

[110] Mathur S., Sharma P., Shukla K.S., Soni P.L. (2013) Potential of tannins for exterior grade plywood. Forest Products Journal 62 (7/8), 559-565.

[111] Younes Shirmohammadli, Davood Efhamisisi, Antonio Pizzi, (2018) Tannins as a sustainable raw material for green chemistry: A review, *Industrial Crops and Products*, Volume 126,2018, Pages 316-332,ISSN 0926-6690, https://doi.org/10.1016/j.indcrop.2018.10.034

[112] De Bruyne, T., Pieters, L., Deelstra, H., & Vlietinck, A. (1999). Condensed vegetable tannins: biodiversity in structure and biological activities. Biochemical Systematics and Ecology, 27(4), 445-459.

[113] Ferreira, D., Brandt, E. V., Coetzee, J., & Malan, E. (1999). Condensed tannins. In Fortschritte der Chemie organischer Naturstoffe/Progress in the Chemistry of Organic Natural Products (pp. 21-67). Springer, Vienna.

[114] Hemingway, R. W., & Karchesy, J. J. (2012). Chemistry and significance of condensed tannins. Springer Science & Business Media.

[115] Schofield, P., Mbugua, D. M., & Pell, A. N. (2001). Analysis of condensed tannins: a review. Animal feed science and technology, 91(1-2), 21-40.

[116] Yoshida, T., Hatano, T., & Ito, H. (2005). Chapter 7, High molecular weight plant poplyphenols (tannins): Prospective functions. In Recent advances in phytochemistry (pp. 163-190). Elsevier Inc.

[117] Das, A. K., Islam, M. N., Faruk, M. O., Ashaduzzaman, M., & Dungani, R. (2020). Review on tannins: Extraction processes, applications and possibilities. South African Journal of Botany, 135, 58-70. Doi:10.1016/j.sajb.2020.08.008

[118] Fraga-Corral M, García-Oliveira P, Pereira AG, Lourenço-Lopes C, Jimenez-Lopez C, Prieto MA, Simal-Gandara (2020) J. Technological Application of Tannin-Based Extracts. Molecules. Jan 30; 25(3):614. DOI: 10.3390/molecules25030614. PMID: 32019231; PMCID: PMC7037717

[119] Pedro L. de Hoyos-Martínez, Juliette Merle, Jalel Labidi, Fatima Charrier – El Bouhtoury (2019) Tannins extraction: A key point for their valorization and cleaner production, Journal of Cleaner Production, Volume 206, Pages 1138-1155, ISSN 0959-6526, https://doi.org/10.1016/j.jclepro.2018.09.243.

[120] Yazaki, Y. (2015). Utilization of flavonoid compounds from bark and wood: a review. *Natural product communications*, 10(30, 5130520. https://doi.org/ 10.1177/1934578X1501000333

[121] Dunky, M. (2020). Wood Adhesives Based on Natural Resources: A Critical Review Part III. Tannin-and Lignin-Based Adhesives. Reviews of Adhesion and Adhesives, 8(4), 379-525. https://doi.org/10.7569/RAA-2020-097313

[122] Elaine C. Ramires, Elisabete Frollini, (2012) Tannin–phenolic resins: Synthesis, characterization, and application as matrix in biobased composites reinforced with sisal fibres, Composites Part B: Engineering, Volume 43, Issue 7, Pages 2851-2860, ISSN 1359-8368, https://doi.org/10.1016/j.compositesb.2012.04.049.

[123] Barbosa Jr, V., Ramires, E. C., Razera, I. A. T., & Frollini, E. (2010). Biobased composites from tannin–phenolic polymers reinforced with coir fibers. Industrial Crops and Products, 32(3), 305-312.

[124] Nicollin, A., Kueny, R., Toniazzo, L., & Pizzi, A. (2012). High density biocomposite from natural fibers and tannin resin. Journal of Adhesion Science and Technology, 26(10-11), 1537-1545.

[125] Nicollin, A., Li, X., Girods, P., Pizzi, A., & Rogaume, Y. (2013). Fast pressing composite using tannin-furfuryl alcohol resin and vegetal fibers reinforcement. Journal of Renewable Materials, 1(4), 311-316.

[126] Ramires, E. C., & Frollini, E. (2012). Tannin–phenolic resins: synthesis, characterization, and application as matrix in biobased composites reinforced with sisal fibers. Composites Part B: Engineering, 43(7), 2851-2860.

[127] Sauget, A., Nicollin, A., & Pizzi, A. (2013). Fabrication and mechanical analysis of mimosa tannin and commercial flax fibers biocomposites. Journal of adhesion science and technology, 27(20), 2204-2218.

[128] Zhu, J., Abhyankar, H., Nassiopoulos, E., & Njuguna, J. (2012). Tannin-based flax fibre reinforced composites for structural applications in vehicles. In IOP Conference Series: Materials Science and Engineering (Vol. 40(1), p. 012030, IOP Publishing.

[129] Zhu, J., Njuguna, J., Abhyankar, H., Zhu, H., Perreux, D., Thiebaud, F., ... & Nicollin, A. (2013). Effect of fibre configurations on mechanical properties of flax/tannin composites. Industrial crops and products, 50, 68-76.

[130] Zhu, J., Zhu, H., Immonen, K., Brighton, J., & Abhyankar, H. (2015). Improving mechanical properties of novel flax/tannin composites through different chemical treatments. Industrial Crops and Products, 67, 346-354.

[131] Zhu, X., Hou, X., Ma, B., Xu, H., & Yang, Y. (2019). Chitosan/gallnut tannins composite fiber with improved tensile, antibacterial and fluorescence properties. Carbohydrate polymers, 226, 115311.

[132] Durrens, P. (1981). The phenoloxidases of the ascomycete Podospora anserina: the three forms of the major laccase activity. Archives of Microbiology, 130(2), 121-124.

[133] Froehner, S. C., & Eriksson, K. E. (1974). Induction of Neurospora crassa laccase with protein synthesis inhibitors. Journal of bacteriology, 120(1), 450-457.

[134] Kurtz, M. B., & Champ, S. P. (1982). Purification and characterization of the conidial laccase of Aspergillus nidulans. Journal of bacteriology, 151(3), 1338-1345.

[135] Dubernet, M., Ribereau-Gayon, P., Lerner, H. R., Harel, E., & Mayer, A. M. (1977). Purification and properties of laccase from Botrytis cinerea. Phytochemistry, 16(2), 191-193.

[136] Gigi, O., Marbach, I., & Mayer, A. M. (1981). Properties of gallic acid-induced extracellular laccase of Botrytis cinerea. Phytochemistry, 20(6), 1211-1213.

[137] Schánĕl, L. (1967). A new polyphenoloxidase test for distinguishing between wood-rotting fungi. Biologia Plantarum, 9(1), 41-48.

[138] Chefetz, B., Chen, Y., & Hadar, Y. (1998). Purification and characterization of laccase fromChaetomium thermophilium and its role in humification. Applied and Environmental Microbiology, 64(9), 3175-3179.

[139] Mayer, A. M. (1986). Polyphenol oxidases in plants-recent progress. Phytochemistry, 26(1), 11-20.

[140] Reinhammar, B. & Malmstrom, B. G. (1981). Metal Ions in Biology: Copper Proteins, 3. *Wiley, New York*, 109-149.

[141] Reinhammar, B. (1970). Purification and properties of laccase and stellacyanin from Rhus vernicifera. *Biochimica et Biophysica Acta (BBA)-Bioenergetics*, *205*(1), 35-47.

[142] Senthivelan, T., Kanagaraj, J., & Panda, R. C. (2016). Recent trends in fungal laccase for various industrial applications: an eco-friendly approach-a review. Biotechnology and Bioprocess Engineering, 21(1), 19-38.

[143] Shekher, R., Sehgal, S., Kamthania, M., & Kumar, A. (2011). Laccase: microbial sources, production, purification, and potential biotechnological applications. Enzyme Research, 1-11.

[144] Yaropolov, A. I., Skorobogat'Ko, O. V., Vartanov, S. S., & Varfolomeyev, S. D. (1994). Laccase. *Applied Biochemistry and Biotechnology*, 49(3), 257-280. https://doi.org/10.1007/BF02783061

[145] Bertrand, B., Martínez-Morales, F., & Trejo-Hernandez, M. R. (2013). Fungal laccases: induction and production. Revista mexicana de ingeniería química, 12(3), 473-488.

[146] Ali, W. B., Chaduli, D., Navarro, D., Lechat, C., Turbé-Doan, A., Bertrand, E., ... & Mechichi, T. (2020). Screening of five marine-derived fungal strains for their potential to produce oxidases with laccase activities suitable for biotechnological applications. BMC biotechnology, 20(1), 1-13.

[147] Debnath, R., Mistry, P., Roy, P., Roy, B., & Saha, T. (2021). Partial purification and characterization of a thermophilic and alkali-stable laccase of Phoma herbarum isolate KU4 with dye-decolorization efficiency. Preparative Biochemistry & Biotechnology, 1-18.

[148] Desai, S. S., Tennali, G. B., Channur, N., Anup, A. C., Deshpande, G., & Murtuza, B. A. (2011). Isolation of laccase producing fungi and partial characterization of laccase. Biotechnol. Bioinf. Bioeng, 1(4), 543-549.

[149] Devasia, S., & Nair, A. J. (2016). Screening of potent laccase producing organisms based on the oxidation pattern of different phenolic substrates. Int J Curr Microbiol App Sci, 5(5), 127-137.

[150] Forootanfar, H., & Faramarzi, M. A. (2015). Insights into laccase producing organisms, fermentation states, purification strategies, and biotechnological applications. Biotechnology progress, 31(6), 1443-1463.

[151] Khalid, N., Asgher, M., & Qamar, S. A. (2020). Evolving trend of Boletus versicolor IBL-04 by chemical mutagenesis to overproduce laccase: Process optimization, 3-step purification, and characterization. Industrial Crops and Products, 155, 112771.

[152] Kiiskinen, L. L., Rättö, M., & Kruus, K. (2004). Screening for novel laccase-producing microbes. Journal of applied microbiology, 97(3), 640-646.

[153] Luterek, J., Gianfreda, L., Wojtas-Wasilewska, M., Rogalski, J., Jaszek, M., Malarczyk, E., ... & Leonowicz, A. (1997). Screening of the wood-rotting fungi for laccase production: induction by ferulic acid, partial purification, and immobilization of laccase from the high laccase-producing strain, Cerrena unicolor. Acta Microbiologica Polonica, 46(3).

[154] Nikolaivits, E., Siaperas, R., Agrafiotis, A., Ouazzani, J., Magoulas, A., Gioti, A., & Topakas, E. (2021). Functional and transcriptomic investigation of laccase activity in the presence of PCB29 identifies two novel enzymes and the multicopper oxidase repertoire of a marine-derived fungus. Science of The Total Environment, 145818.

[155] Revankar, M. S., & Lele, S. S. (2006). Enhanced production of laccase using a new isolate of white rot fungus WR-1. Process Biochemistry, 41(3), 581-588.

[156] Tapia-Tussell, R., Pérez-Brito, D., Rojas-Herrera, R., Cortes-Velazquez, A., Rivera-Muñoz, G., & Solis-Pereira, S. (2011). New laccase-producing fungi isolates with biotechnological potential in dye decolorization. African Journal of Biotechnology, 10(50), 10134-10142.

[157] Wang, F., Xu, L., Zhao, L., Ding, Z., Ma, H., & Terry, N. (2019). Fungal laccase production from lignocellulosic agricultural wastes by solid-state fermentation: a review. Microorganisms, 7(12), 665.

[158] Xu, F., Shin, W., Brown, S. H., Wahleithner, J. A., Sundaram, U. M., & Solomon, E. I. (1996). A study of a series of recombinant fungal laccases and bilirubin oxidase that exhibit significant differences in redox potential, substrate specificity, and stability. Biochimica et Biophysica Acta (BBA)-Protein Structure and Molecular Enzymology, 1292(2), 303-311. https://doi.org/10.1016/0167-4838(95)00210-3

[159] Yang, X., Wu, Y., Zhang, Y., Yang, E., Qu, Y., Xu, H., ... & Yan, J. (2020). A thermo-active laccase isoenzyme from Trametes trogii and its potential for dye decolorization at high temperature. Frontiers in microbiology, 11, 241.

[160] Bauer, C. G., Kühn, A., Gajovic, N., Skorobogatko, O., Holt, P. J., Bruce, N. C., ... & Scheller, F. W. (1999). New enzyme sensors for morphine and codeine based on morphine dehydrogenase and laccase. Fresenius' journal of analytical chemistry, 364(1), 179-183. DOI: 10.1007/s002160051320

[161] Haars A. & Hüttermann A (1980). Macromolecular mechanism of lignin degradation by Fomes annosus. Naturwissenschaften 67, 39-40.

[162] Haars, A. & Hüttermann, A. (1980). Function of laccase in the white-rot fungus Fomes annosus. Archives of microbiology, 125(3), 233-237.

[163] Mayer, A. M., & Staples, R. C. (2002). Laccase: new functions for an old enzyme. Phytochemistry, 60(6), 551-565. https://doi.org/10.1016/S0031-9422(02)00171-1

[164] Moreno, A. D., Ibarra, D., Fernández, J. L., & Ballesteros, M. (2012). Different laccase detoxification strategies for ethanol production from lignocellulosic biomass by the thermotolerant yeast Kluyveromyces marxianus CECT 10875. Bioresource Technology, 106, 101-109. https://doi.org/10.1016/j.biortech.2011.11.108

[165] Patel, N., Shahane, S., Majumdar, R., & Mishra, U. (2019). Mode of action, properties, production, and

application of laccase: a review. Recent patents on biotechnology, 13(1), 19-32. https://doi.org/10.2174/1872208312666180821161015

[166] Pezzella, C., Guarino, L., & Piscitelli, A. (2015). How to enjoy laccases. *Cellular and Molecular Life Sciences*, *72*(5), 923-940.

[167] Plácido, J., & Capareda, S. (2015). Ligninolytic enzymes: a biotechnological alternative for bioethanol production. *Bioresources and Bioprocessing*, *2*(1), 1-12. https://doi.org/10.1186/s40643-015-0049-5

[168] Sealey, J. E., Ragauskas, A. J., & Elder, T. J. (1999). Investigations into laccase-mediator delignification of kraft pulps. Holzforschung, 53, 498-502.

[169] Tanaka, H., Itakura, S., & Enoki, A. (1999). Hydroxyl radical generation by an extracellular low-molecular-weight substance and phenol oxidase activity during wood degradation by the white-rot basidiomycete Trametes versicolor. *Journal of Biotechnology*, 75(1), 57-70.

[170] Widsten, P., & Kandelbauer, A. (2008). Laccase applications in the forest products industry: a review. Enzyme and microbial technology, 42(4), 293-307.

[171] Eriksson, K. E. L., Blanchette, R. A., & Ander, P. (1990). Biodegradation of lignin. In Microbial and enzymatic degradation of wood and wood components (pp. 225-333). Springer, Berlin, Heidelberg.

[172] Kawai, S., Umezawa, T., & Higuchi, T. (1988). Degradation mechanisms of phenolic β-1 lignin substructure model compounds by laccase of Coriolus versicolor. Archives of biochemistry and biophysics, 262(1), 99-110. https://doi.org/10.1016/0003-9861(88)90172-5

[173] Call, H. P., & Mücke, I. (1997). History, overview and applications of mediated lignolytic systems, especially laccase-mediator-systems (Lignozym®-process). Journal of biotechnology, 53(2-3), 163-202. https://doi.org/10.1016/S0168-1656(97)01683-0

[174] Cullen, D. (1997). Recent advances on the molecular genetics of ligninolytic fungi. Journal of Biotechnology, 53(2-3), 273-289.

[175] Kersten, P. J., & Kirk, T. K. (1987). Involvement of a new enzyme, glyoxal oxidase, in extracellular H2O2 production by Phanerochaete chrysosporium. Journal of bacteriology, 169(5), 2195-2201.

[176] Kersten, P. J., Tien, M., Kalyanaraman, B., & Kirk, T. K. (1985). The ligninase of Phanerochaete chrysosporium generates cation radicals from methoxybenzenes. Journal of Biological Chemistry, 260(5), 2609-2612.

[177] Hammel, K. E., Jensen Jr, K. A., Mozuch, M. D., Landucci, L. L., Tien, M., & Pease, E. A. (1993). Ligninolysis by a purified lignin peroxidase. Journal of Biological Chemistry, 268(17), 12274-12281.

[178] Balakshin, M., Capanema, E., Chen, C. L., Gratzl, J., Kirkman, A., & Gracz, H. (2001a). Biobleaching of pulp with dioxygen in the laccase-mediator system—reaction mechanisms for degradation of residual lignin. Journal of Molecular Catalysis B: Enzymatic, 13(1-3), 1-16. https://doi.org/10.1016/S1381-1177(00)00225-3

[179] Balakshin, M., Chen, C. L., Gratzl, J. S., Kirkman, A. G., & Jakob, H. (2001b). Biobleaching of pulp with dioxygen in laccase-mediator system—effect of variables on the reaction kinetics. Journal of Molecular Catalysis B: Enzymatic, 16(3-4), 205-215.

[180] Balakshin, M., Chen, C. L., Gratzl, J. S., Kirkman, A. G., Jakob, H., & Degussa, A. G. (1999). Biobleaching of pulp with dioxygen in the laccase-mediator system. Holzforschung, 54, 390-396.

[181] Bourbonnais, R., & Paice, M. G. (1992). Demethylation and delignification of kraft pulp by Trametes versicolor laccase in the presence of 2, 2′-azinobis-(3-ethylbenzthiazoline-6-sulphonate). Applied Microbiology and biotechnology, 36(6), 823-827. https://doi.org/10.1007/BF00172202

[182] Bourbonnaus, R. (1996). Enzymatic delignification of kraft pulp using laccase and a mediator. TAPPI J., 79, 199-204.

[183] Majcherczyk, A., Johannes, C., & Hüttermann, A. (1999). Oxidation of aromatic alcohols by laccase from Trametes versicolor mediated by the 2, 2′-azino-bis-(3-ethylbenzothiazoline-6-sulphonic acid) cation radical and dication. Applied microbiology and biotechnology, 51(2), 267-276. https://doi.org/10.1007/s002530051392

[184] Balakshin, M. Y., Chen, C. L., Gratzl, J. S., Kirkman, A. G., Jakob, H., & Degussa, A. G. (2000). Kinetic studies on oxidation of veratryl alcohol by laccase-mediator system. Holzforschung, 54(2), 165-170.

[185] Freudenreich, J., Amann, M., Fritz-Langhals, E., & Stohrer, J. (1998). Understanding the Lignozym-process. In International Pulp Bleaching Conference Vol.1, 71-76.

[186] Munk, L., Sitarz, A. K., Kalyani, D. C., Mikkelsen, J. D., & Meyer, A. S. (2015). Can laccases catalyze bond cleavage in lignin? Biotechnology advances, 33(1), 13-24. https://doi.org/10.1016/j.biotechadv.2014.12.008

[187] Potthast, A., Rosenau, T., & Fischer, K. (2001). Oxidation of benzyl alcohols by the laccase-mediator system (LMS) a comprehensive kinetic description. *Holzforschung* 55, 47-56.

[188] Berthet, S., Thevenin, J., Baratiny, D., Demont-Caulet, N., Debeaujon, I., Bidzinski, P., & Jouanin, L. (2012). Role of plant laccases in lignin polymerization. Advances in Botanical Research, 61, 145-172. https://doi.org/10.1016/B978-0-12-416023-1.00005-7

[189] De Gonzalo, G., Colpa, D. I., Habib, M. H., & Fraaije, M. W. (2016). Bacterial enzymes involved in lignin degradation. Journal of biotechnology, 236, 110-119. https://doi.org/10.1016/j.jbiotec.2016.08.011

[190] Dean, J. F., & Eriksson, K. E. L. (1994). Laccase and the deposition of lignin in vascular plants. Holzforschung, 48, 21-21.

[191] Felby, C., Pedersen, L. S., & Nielsen, B. R. (1997). Enhanced auto adhesion of wood fibres using phenol oxidases. Holzforschung-International Journal of the Biology, Chemistry, Physics and Technology of Wood, 51(3), 281-286.

[192] Gavnholt, B., & Larsen, K. (2002). Molecular biology of plant laccases in relation to lignin formation. Physiologia plantarum, 116(3), 273-280. https://doi.org/10.1034/j.1399-3054.2002.1160301.x

[193] Ikeda, R., Sugihara, J., Uyama, H., & Kobayashi, S. (1996). Enzymatic oxidative polymerization of 2, 6-dimethylphenol. Macromolecules, 29(27), 8702-8705. https://doi.org/10.1021/ma961055h

[194] Mattinen, M. L., Suortti, T., Gosselink, R. J. A., Argyropoulos, D. S., Evtuguin, D., Suurnäkki, A., & Tamminen, T. (2008). Polymerization of different lignins by laccase. BioResources, 3(2), 549-565.

[195] Tobimatsu, Y., & Schuetz, M. (2019). Lignin polymerization: how do plants manage the chemistry so well? *Current Opinion in Biotechnology*, 56, 75-81. https://doi.org/10.1016/j.copbio.2018.10.001

[196] Zhao, Q., Nakashima, J., Chen, F., Yin, Y., Fu, C., Yun, J., ... & Dixon, R. A. (2013). Laccase is necessary and nonredundant with peroxidase for lignin polymerization during vascular development in Arabidopsis. *The Plant Cell*, 25(10), 3976-3987. https://doi.org/10.1105/tpc.113.117770

[197] Milstein, O., Hüttermann, A., Fründ, R., & Lüdemann, H. D. (1994). Enzymatic co-polymerization of lignin with low-molecular mass compounds. Applied Microbiology and Biotechnology, 40(5), 760-767.

[198] Higuchi, T. (2012a). Biochemistry and molecular biology of wood. Springer Science & Business Media.

[199] Higuchi, T. (Ed.). (2012b). Biosynthesis and biodegradation of wood components. Elsevier.

[200] Kharazipour, A., Bergmann, K., Nonninger, K., & Hüttermann, A. (1998). Properties of fibre boards obtained by activation of the middle lamella lignin of wood fibres with peroxidase and H2O2 before conventional pressing. Journal of adhesion science and technology, 12(10), 1045-1053.

[201] Kharazipour, A., Mai, C., & Hüttermann, A. (1998). Polyphenoles for compounded materials. Polymer degradation and stability, 59(1-3), 237-243.

[202] Nimz, H., Razvi, A., Marquharab, I., & Clad, D. (1972). Bindemittel bzw Klebemittel zur Herstellung von Holzwerkstoffen sowie zur Verklebung von Werkstoffen verschiedener Art. German patent DOS, 2221353.

[203] Hüttermann, A., Mai, C., & Kharazipour, A. (2001). Modification of lignin for the production of new compounded materials. Applied microbiology and biotechnology, 55(4), 387-394.

[204] Kharazipour, A., Hüttermann, A., & Luedemann, H. D. (1997). Enzymatic activation of wood fibres as a means for the production of wood composites. Journal of Adhesion Science and Technology, 11(3), 419-427.

[205] Nasir, M., Gupta, A., Beg, M. D. H., Chua, G. K., Jawaid, M., Kumar, A., & Khan, T. A. (2013). Fabricating eco-friendly binderless fiberboard from laccase-treated rubber wood fiber. BioResources, 8(3), 3599-3608.

[206] Sharma P. & Huttermann A. (2001) Wood composites by Enzymatic methods. Report submitted to Forstbotanisches Institut Der Universitat Gottingen Busgenweg-237077 Gottingen Germany.

[207] Xing, Q., Eadula, S. R., & Lvov, Y. M. (2007). Cellulose fiber enzyme composites fabricated through layer-by-layer nanoassembly. Biomacromolecules, 8(6), 1987-1991.

[208] Nasir, M., Hashim, R., Sulaiman, O., Nordin, N. A., Lamaming, J., & Asim, M. (2015). Laccase, an emerging tool to fabricate green composites: a review. BioResources, 10(3), 6262-6284.

[209] Li, Y., & Pickering, K. L. (2008). Hemp fibre reinforced composites using chelator and enzyme treatments. Composites science and technology, 68(15-16), 3293-3298. https://doi.org/10.1016/j.compscitech.2008.08.022

[210] Bledzki, A. K., Mamun, A. A., Jaszkiewicz, A., & Erdmann, K. (2010). Polypropylene composites with enzyme modified abaca fibre. Composites Science and Technology, 70(5), 854-860.

[211] Peng, X., Zhong, L., Ren, J., & Sun, R. (2010). Laccase and alkali treatments of cellulose fibre: Surface lignin and its influences on fibre surface properties and interfacial behaviour of sisal fibre/ phenolic resin composites. Composites Part A: Applied Science and Manufacturing, 41(12), 1848-1856.

[212] Islam, M. R., Beg, M. D., & Gupta, A. (2013). Characterization of laccase-treated kenaf fibre reinforced recycled polypropylene composites. BioResources, 8(3), 3753-3770.

[213] Dong, A., Yu, Y., Yuan, J., Wang, Q., & Fan, X. (2014). Hydrophobic modification of jute fiber used for composite reinforcement via laccase-mediated grafting. Applied surface science, 301, 418-427.

[214] Vishnu Vardhini, K. J., & Murugan, R. (2017). Effect of laccase and xylanase enzyme treatment on chemical and mechanical properties of banana fiber. Journal of Natural Fibres, 14(2), 217-227. https://doi. org/10.1177/1528083717714479

[215] Vishnu Vardhini, K. J., Murugan, R., & Surjit, R. (2018). Effect of alkali and enzymatic treatments of banana fibre on properties of banana/polypropylene composites. Journal of Industrial Textiles, 47(7), 1849-1864. https://doi. org/10.1177/1528083717714479

[216] Thakur, K., Kalia, S., Kaith, B. S., Pathania, D., & Kumar, A. (2015). Surface functionalization of coconut fibres by enzymatic biografting of syringaldehyde for the development of biocomposites. *RSC advances*, 5(94), 76844-76851. https://doi.org/10.1039/ C5RA14891J

[217] Nasir, M., Hashim, R., Sulaiman, O., Gupta, A., Khan, T. A., Jawaid, M., & Asim, M. (2017). Natural fiber improvement by laccase; optimization, characterization and application in medium density fiberboard. Journal of Natural Fibres, 14(3), 379-389.

[218] Dong, A., Li, F., Fan, X., Wang, Q., Yu, Y., Wang, P., & Cavaco-Paulo, A. (2018). Enzymatic modification of jute fabrics for enhancing the reinforcement in jute/PP composites. Journal of Thermoplastic Composite Materials, 31(4), 483-499.

[219] Singh, A. & Palsule, S. (2016). Jute fiber reinforced chemically functionalized polypropylene self-compatibilizing composites by Palsule process. Journal of Composite Materials, 50(9), 1199-1212.

[220] Brodowsky, H. M., Hennig, A., Müller, M. T., Werner, A., Zhandarov, S., & Gohs, U. (2020). Laccase-Enzyme Treated Flax Fibre for Use in Natural Fibre Epoxy Composites. Materials, 13(20), 4529.

[221] Liu, M., Baum, A., Odermatt, J., Berger, J., Yu, L., Zeuner, B., Thygesen, A., Holck, J. and (2017). Oxidation of lignin in hemp fibres by laccase: Effects on mechanical properties of hemp fibres and unidirectional fibre/epoxy composites. Composites Part A: Applied Science and Manufacturing, 95, 377-387.

[222] Acero, E. H., Kudanga, T., Ortner, A., Kaluzna, I., De Wildeman, S., Nyanhongo, G. S., & Guebitz, G. M. (2014). Laccase functionalization of flax and coconut fibers. Polymers, 6(6), 1676-1684. https://doi.org/10.3390/ polym6061676

[223] Kalia, S., Thakur, K., Kumar, A., & Celli, A. (2014). Laccase-assisted surface functionalization of lignocellulosics. Journal of Molecular Catalysis B: Enzymatic, 102, 48-58. https://doi.org/10.1016/j. molcatb.2014.01.014

[224] Mayer, A. M. (2006). Polyphenol oxidases in plants and fungi: going places? A review. Phytochemistry,

67(21), 2318-2331. https://doi.org/10.1016/j.phytochem. 2006.08.006

[225] Nasir M., Asim M., Singh K. (2021) Fiberboard Manufacturing from Laccase Activated Lignin Based Bioadhesive. In: Jawaid M., Khan T.A., Nasir M., Asim M. (eds) Eco-Friendly Adhesives for Wood and Natural Fiber Composites. Composites Science and Technology. Springer, Singapore. https://doi.org/10.1007/978-981-33-4749-6_3

Chapter 3

Functionality Based Design of Sustainable Bio-Composite

MD Rajbanul Akhond and Ahmed Sharif

Abstract

Bio-composites have diverse functional demands for many structural, electrical, electronic, and medical applications. An expansion of the composite functionality is achieved by manipulating the material and design scheme. Smart selection of matrix-reinforcement combinations will lead to applications that have never even been considered. Research holds a huge potential to create a wide variety of usable materials by mixing different fillers and modifying the parameters. Apart from selecting the polymer and the filler, the engineer will have to understand the compatibility of the polymer and the filler, dispersion, and bonding behavior making the design of polymer nanocomposite a rather complex system. In this chapter, we have tried to display different functional materials development pursuit.

Keywords: bio-composite, functional design, conductive polymer, nanotube encapsulation, bioactive, biodegradable, food packaging, medical applications

1. Introduction

Biocomposite materials derived from biodegradable, renewable sources have perceived substantial attention in current times, in particular due to the amplified consciousness towards more environmentally sustainable technologies. In most of the cases bio-composites offer weight reduction, supplemented capabilities (e.g., structural, electrical, electronic, and medical) and occupational health benefits. These bio-composites have great potential for their use in electronic, automobile, packaging, insulations and construction industries with a potential to disassociate material costs from the fluctuating price of oil and energy.

2. Design with functionality approach

Addition of reinforcements to a bio-composite matrix does more than just enhance on the properties of the composite but also imparts some other "unexpected" properties that can satisfy these requirements. For example, bio-composites with conductive polymer matrices impart unique functional properties which makes them useful for medical applications. On the other hand, use of clay nanotubes as reinforcement materials not only enhances mechanical properties but also opens the way for surface topology enhancements resulting more unique applications. Moreover, smart choice of matrix-reinforcement combinations and understanding their surface interactions can result in applications never even thought about. Such as, keeping both the in-situ surface interactions and interactions with

IntechOpen

the environment in the mind, choice of appropriate matrix-reinforcement combination can reward us with bio-degradable food packaging materials. This creates a situation where the excellent design of bio-composites provides an extension of the composite functionality. By combining the filler and manipulating the parameter space, bio-composite research holds a vast possibility to produce a wide range of functional materials with controllable properties. To illustrate functionality-based designs, a few of the bio-composites that have been formulated with functional properties are presented here.

2.1 Conductive polymer bio-composites

There is an emerging attention in the field of natural fiber fortified conductive polymer combinations in numerous aspects of technical functions. Such attention ascends from the fact that the natural fiber supported polymer matrix composites show conjoined possessions of useful insulation and elevated degree of anticipated mechanical strength that allows it to be magnificent structural provision for conductors. Such exclusive characteristics allow these composite structures to be exploited in broadened regions for instance terminals, connectors, switches, insulators, printed circuit boards, industrial and house hold plugs, panels and so on [1, 2]. Moreover, the dielectric features of materials greatly control the transformation of the electromagnetic energy toward heat [3]. At present, several textile fibers and fabrics such as cotton, nylon, lycra, polyester, viscose rayon, and wool, are being employed with technical polymers for diverse functions e.g. conductive textiles, heating appliances, electro-magnetic interference, super-capacitor, shielding and antimicrobial cloths and so on [4–6]. Nonetheless, conducting polymers with receptive to electrical stimulation are competent to bio- functions such as tissue scaffolds for the restoration or replacement of damaged or malfunctioning tissues [7]. A graphic demonstration showing the significance of designing biomimetic conducting polymer-based materials is exhibited in **Figure 1**. A continuing push to foster more bio-conforming and intrinsically ecological conductive polymer types with various functionalization techniques are evolved for different possible applications. Optimization of various properties (e.g., conductivity, roughness, porosity, hydrophobicity and degradability) in combination with the attachment of biological molecules have made the conductive polymers as a promising candidate

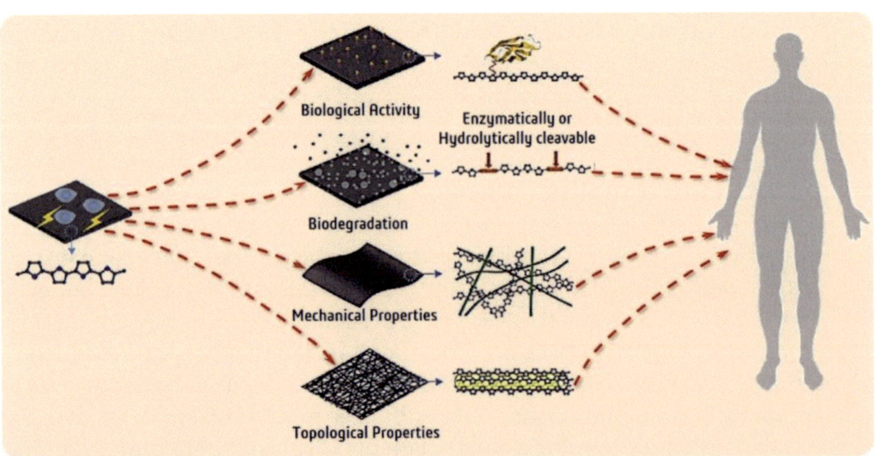

Figure 1.
Schematic illustration of the critical considered aspects for designing biomimetic conducting polymer-based materials. Redrawn from [7].

for biomedical applications. This binding of organic molecules can be conducted via four key synthetic ways [8] (**Figure 2**). First process is adsorption where a solution of functionalizing chemical is put together with the synthesized polymer. The organic molecules are attached to the polymer structure due to the stable contacts with ionic interaction [9]. Secondly, engrossing the functional molecules within the polymer matrix. The entrapment is usually carried out by combining the matrix monomer and functionalizing chemical agent with the polymerization reaction afterwards [10]. Thirdly, by cementing the biomolecules to the polymer matrix with covalent bonding. In this case, functionalization of the polymer is more stable and long-lasting [8]. Lastly by doping practice with a wide variety of charged molecules for inducing the polymers to conductive character [8, 11]. The obvious benefit of conductive polymers is their immense flexibility in terms of dopant selection for proper functionalization to suit a particular usage [12]. Therefore, the fundamental understanding of these dielectric properties is essential for various industrial applications.

Conductive polymers have more extensive favorable circumstances over other electro-active biomaterials (for example electrets, piezoelectric and photovoltaic materials) according to electrical perspective [13]. They have astounding authority over electrical stimulus, have a high conductivity to weight ratio, can have generally excellent electrical just as optical properties, and can likewise add to making biodegradable, permeable, and biocompatible items [14–17]. Their physical, synthetic, and electrical properties can be custom fitted for providing explicit necessities of their applications and viewed as one of their exceptional focal points. This can be possible by fusing antibodies, proteins and other organic moieties [14, 16, 18]. In addition, such helpful properties of conductive polymers can be controlled, and changed even after combination through incitement (for example using various methods like electricity, light, pH and so forth.) [19–21].

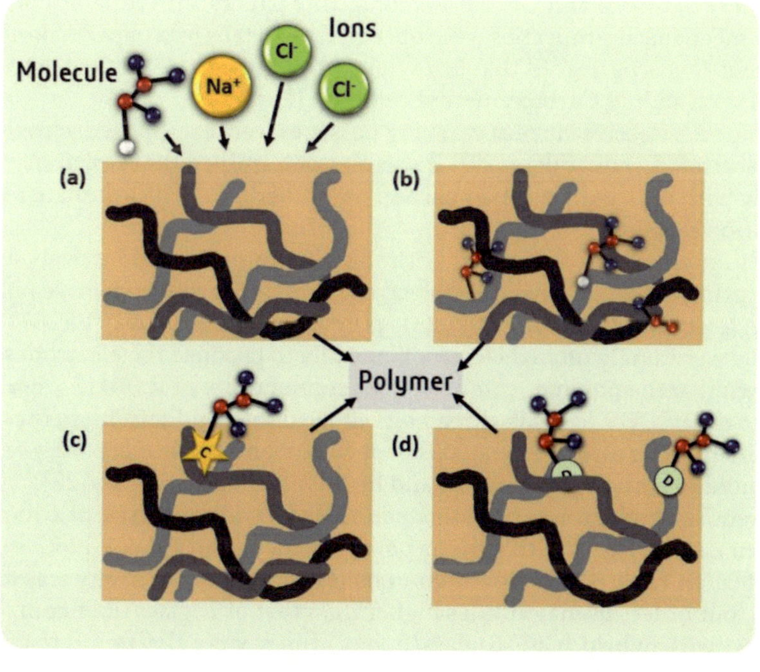

Figure 2.
Methods of functionalizing conductive polymers: (a) physical absorption, (b) entrapping, (c) covalent bonding and (d) exploiting the doping mechanism.

Recently, their unexpected applications in functional papers just as packaging industries have drawn exceptional consideration. A few investigations affirmed critical need of utilizing such conductive polymers in electrical applications. Coated paper with conducting properties can be utilized to produce anti-static and electromagnetic shielding papers, anti-bacterial papers, novel wall coverings and electrical resistive heating papers [22–24]. Johnston's group designed conducing paper using natural strands and conductive polymers, where unbleached bagasse as well as rice straw filaments were injected into polyaniline (PANi). Results demonstrated that increased conductivity was achieved with the increase of PANi in the composite [22]. Youssef et al. also architectured a composite system based on unbleached cellulosic bagasse or rice straw fibers and PANi as conducting polymer by means of emulsion polymerization. In this context, PANi was formulated with in situ polymerization via oxidative reaction with help of ammonium persulfate accompanied with a specific quantity of emulsifier n-dodecylbenzenesulfonic acid (DBSA) and dopant HCl. DBSA was used as emulsifier to catalyze the polymerization reaction of aniline via cation-radical mechanism in the presence of unbleached pulp fibers and also to increase the electrical conductivity of the formed PANi/pulp fiber composites.

In addition, Pramila Devi's group have studied the cure characteristics, thermal and microwave properties, DC conductivity, and mechanical properties of both natural rubber (NR)/polypyrrole and natural rubber/polypyrrole/polypyrrole-coated short nylon fiber composite. The composites were designed by a two-step process: in situ polymerization in NR latex followed by compounding in a two-roll mill. It was noted that the natural rubber/polypyrrole composite's DC conductivity was only improved at very high polypyrrole load and a maximum conductivity of 8.3×10^{-4} S/cm was reached at loading time of 100 phr [25].

In addition to this, Jabbour's group developed conductive papers of graphite particles (GPs)/ carbon fibers (CFs)/cellulose fibers (FBs) composites with low cost of production, good mechanical properties, and tunable electrical conductivity. Flexible GP/FB, CF/FB and GP/CF/FB composite papers with tunable conductivity and good mechanical properties were obtained by means of a papermaking production process. It was perceived that ultraviolet absorbance in UV–vis spectroscopy amplified with adding Carboxymethyl cellulose [26].

In addition, effective current carrying passages were established across agglomerates or scattered nano-fillers [27]. It was also determined that resistivity varied drastically with the distribution of carbon nanofiber into polycarbonate regulated by sonication facilities.

In addition, conductive polymer with silk fiber bunch was effectively used in the manufacturing of thread-fashioned electrodes. In this case, conductive polyelectrolyte, poly(3,4-ethylenedioxythiophene) -poly(styrene sulfonate) (PEDOT-PSS) and silk bundle were amalgamated electrochemically to produce the electrodes. It has been shown that the polymer composite has conductivity of 0.00117 S/cm. Adding glycerol to the PEDOT-PSS silk thread has allowed the conductivity to increase to 0.102 S/cm. It has also been stated that such biocompatible electrodes can be implemented in both the biomedical and health promotion sectors [28].

Different research groups investigated the dielectric properties of composites made from natural fibers with various polymers including the conductive ones. Chicken feather fiber (CFF)/Epoxy composites' electrical resistivity was found to be two to four orders of magnitudes higher than that of E-glass fiber composites. Composites with hybrid (CFF-E-glass/Epoxy) fibers were also found to typically have a low dielectric constant value [29].

On the other hand, Xia and Lu fabricated highly conductive polymers with composite silk fibroin fibers through in-situ polymerization. Polypyrrole/silk fibroin, polyaniline/silk fibroin, and poly3,4-ethylene-dioxythiophene/silk fibroin composite fibroins were reported to exhibit differed conductivity in the range of $3.8-4.2\times10^{-1}$, $0.9-1.2\times10^{-2}$ and $4.9-5.2\times10^{-3}$ Scm^{-1}, respectively. It has also been exhibited that these composites confirmed better electrical and thermal characteristics showing possibility of uses in textile and biological areas as novel functional materials [30].

Gelfuso's group also studied the electrical resistivity of composites based on composites made from polypropylene/coconut fibers. They aimed to investigate the electrical properties of low-cost and composites that are environmentally friendly in order to enhance their implementation in industrial applications [31].

Moreover, W. Jia's group researched and analyzed the electrical conductivity of composites based on epoxy resin with polyaniline-dodecylbenzenesulfonic acid (DBSA) fillers. They used both conductive filler PANI-DBSA in powder and paste forming the composite with matrix polymer bisphenol, hydride hardener, and epoxy resin as well as accelerator. Results showed a conductivity of the order 10^{-3} at high filler content [32].

In addition, Wang's group had introduced the theory of percolation and its principles by conducting experiments to study the effect of absorption of moisture on electrical conductivity in natural plastic composites. It was confirmed that there is no observable electrical conductivity in the dry natural fiber-reinforced polymer composite. But electrical conductivity for the natural fiber composite was achieved successfully after water submersion [33].

Moreover, there was investigation of the dielectric properties of date palm fiber/epoxy composite where three relaxation processes were found. These processes were primarily relaxation in the α mode, relaxation due to diffusion of carrier charges for high temperatures above glass transition and low frequencies, and relaxation in the interfaces or Maxwell -Wagner-Sillars relaxation [34–36].

Similarly, as a biodegradable material, conductive nanocomposite using polypyrrole/dextrin was synthesized. The conductivity of the nanocomposites was investigated using four probe methods and analyzed with 2,2-diphenyl-1-picrylhydrazyl assay (DPPH) for antioxidant activity. It was shown that by increasing the polypyrrole in the matrix, both the conductivity and antioxidant activity have improved. The nanocomposites were indicated to be considerably effective against all such bacteria studied. It was also stated that composites in the range of 30.18–74.52% degradation is biodegradable in the natural environment [37].

In addition, Pavlović's group studied the effects of electrical conductivity of electrodeposited copper powder material filled into the lignocellulose matrix. The conductivity measurements were shown to have S-shaped dependence with a percolation transition from non-conductive to conductive region. The concentration percolation threshold was reached at a copper fraction of 14.4% (v/v) volume [38].

2.2 Clay nanotube encapsulation for functional bio-composites

Halloysite clay is a natural medium of tubules formed by rolled kaolin sheets. Halloysite is alumosilicate and is chemically identical to kaolin although in some position of aluminum usually, contains a small amount of metal ions replacing it. Usually, 10–15 layers of alumosilicate roll into the cylinder and its wall packing can be monitored with 0.72 nm X-ray reflection (001) for dry halloysite [39–42]. Halloysite tube surface is silica, and its innermost surface is alumina, having a good negative zeta-potential of ca. -30 mV on the surface of the tube and + 25 mV on the innermost tube in aqueous dispersions at standard pH. The diameter of the halloysite tubes is 40-70 nm with an inner lumen diameter of 10-15 nm and a length

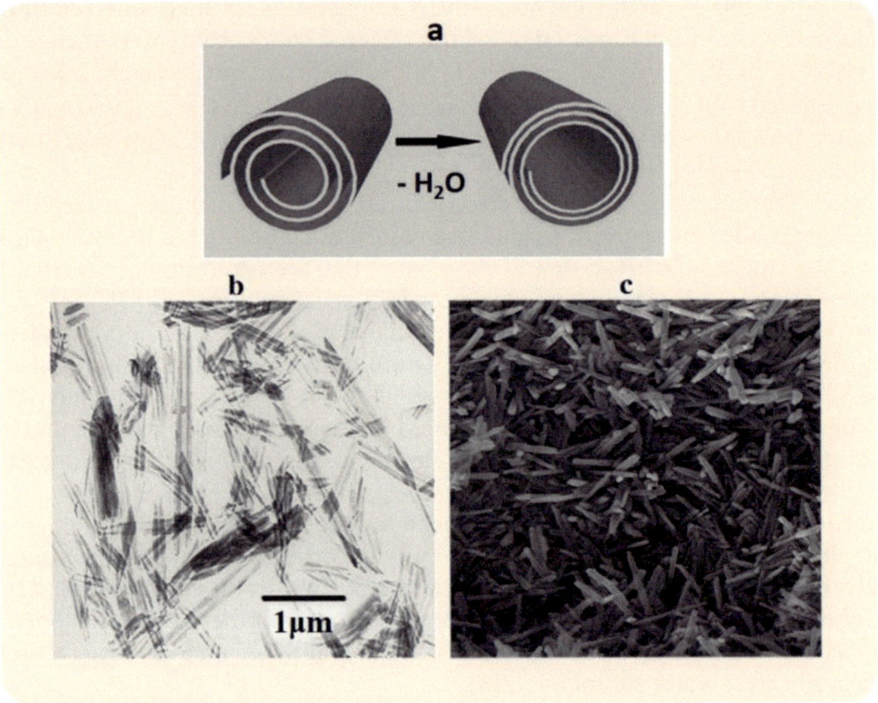

Figure 3.
Scheme of tightening halloysite tube walls during heating dehydration (a). Aqueous dispersion TEM and dry powder SEM images of halloysite from Dragon Mine, Utah, Applied Minerals Inc. (b–c). Redrawn from [43].

of 1500 ± 500 nm (**Figure 3**) [42, 44, 45]. It is interesting that sonication leads to the formation of halloysite-like tubes over a long time of kaolin aqueous dispersion, although the concentration of the substance is very low and the tube shape is rather poor [46].

In contrast with platy clays such as montmorillonite, kaolin and laponite packed in larger crystallites, halloysite has an important advantage that these nanotubes do not need exfoliation and can easily be dispersed in water or polar polymers. Halloysite-water dispersion is stable for a few hours and can be re-dispersed with quick shaking or brief sonication. It was proven that nano clay-tubes have a solid interfacial characteristic with biopolymers (e.g., polysaccharides, polyamides) and also with polar polymers (e.g., polyacrylates, epoxy, polyvinylchloride, polyethylene). Most importantly, these nanotubes can be incorporated into polymer matrix with proper dispersion. These clay-based nanotubes develop a sort of "structural framework" in bulk polymers which enhances the strength the composite. In addition, these "skeleton bones" can be filled with bioactive substances, such as loading real bones with a marrow offering additional functionality. Halloysite is biocompatible "green" material and its simple processing combined with low cost makes it a prospect for polymeric composites in nano-architecture.

The straightforward application of clay nanotubes is their use in the modification of biological devices' surface topography, such as microfluidic devices, to distinguish the circulating tumor cells from non-malignant. Halloysite nanotubes functionalized with human selectin protein have been successfully used to isolate myeloid leukemia from colon cancer cells [47]. The system was based on internal-coated 300 μm glass tubes with P or E-selectin molecules and alternated with negatively charged halloysite nanotubes with poly-L -lysine. Microscopic experiments showed that halloysite nanotubes adapted to the glass surface greatly decreased the speed of rolling and increased the number of cells captured. Improving cell separation is caused by the increase in the total surface area following the immobilization of clay nanotubes and the resulting increase in the amount of selectin molecules responsible for the cell capture. In some other situations, the same method can be used where we can pick a suitable molecular agent (i.e., antibodies, receptors, and aptamers), and then exclusively remove the cells accordingly.

In addition, it was found that the nano clay tubes could be functionalized with sensory stimulus molecules and could be stuffed with gradually liberating molecules, enhancing the coating functionality. It was also established that enriched human dermal fibroblast addition on halloysite nanotube coatings which spread, proliferate and sustain the cellular phenotype [48].

An emerging application of halloysite nanotubes in biomaterials is their use in tissue and dental engineering scaffolds. Bottino's group investigated the manufacture of three-dimensional endodontic regenerative scaffolds based on electro spun polymers doped with halloysite nanotubes. Nanotubes were found predominantly inside the polydioxanone fibers and allowed an increase in the diameter of the fiber. Halloysite addition also impacted the mechanical properties of scaffolds. It has been shown that fibroblast cells derived from human dental pulp exhibited the usual proliferation rate, indicating high biocompatibility of polymer scaffolds doped with electro spun halloysite [49].

Liu's group have used the idea of using halloysite nanotubes as dopants in the fabrication of chitosan-based tissue engineering scaffolds. Nanocomposite halloysite-doped scaffolds demonstrated the enhanced mechanical and thermal properties of compressive power, compressive assembly, and temperature stability compared to pure chitosan scaffolds. Human fibroblasts had effectively colonized these scaffolds [50].

Another fascinating field of research is the tailoring of nanomaterials to cell surfaces that enables living microbial cells to make new functionalities. Max Planck Institute's team led by Helmuth Möhwald proposed layer-by-layer (LbL) encapsulation of biological cells and it was first demonstrated for surface modification of *Escherichia coli* bacteria and erythrocytes (red blood cells) [51, 52]. These so-called "cyborg cells" have recently become a common and promising biomaterial for composites [53–58]. For example, living cells can be anchored with surface-attached magnetic nanoparticles by means of micro-capsules that enable in vitro manipulation through a magnetic field. Halloysite nanotubes were also immobilized on microbial cell walls by LbL deposition with opposite charged polymers [54, 58].

Halloysite-coated cells can be used as adaptable patterns for the manufacture of hollow inorganic microcapsules by calcination.; On the other hand, the fascinating uses are anticipated from live cells carrying a vehicle of halloysite nanotubes packed with different ingredients in advance of the confinement process. These "nanocarrier" cells could be considered as an active composite structure that provides a load of nutrients, defensive antibodies or biocides, DNA, and enzymes for the cells. For example, the extended release of glucose loaded into halloysite nanotubes immobilized on yeast cells has been demonstrated as a proof of principle [57].

2.3 Functional bio-composites for food packaging

Packaging must ensure its fundamental function as container and food damage against physical damage, by displaying adequate mechanical properties, but also by regulating the properties of mass transfer to restrict food degradation reactions. It has been shown that lignocellulosic fibers frequently act as defects that affect both mechanical and permeability properties, degrading the former and increasing the latter. Increasing material permeability by choosing the right formulation seems to be a valuable approach, especially for the packaging of respiring products such as fruit and vegetables.

Mechanical properties for multiple combinations of bioplastics and lignocellulose fibers were widely explored and published. Despite the value of the properties of mass transit, however, the state of information on this subject remains very weak. Clearly, a need for fundamental research turned towards full-bio-composites emerges from this context, by developing understanding and modeling methods capable of considering the entire intricacy of the systems. The present blockages are the deficiency of proper know-how on the underlying characteristics of vegetable yarns in mass transfer and the function of the interphase between the fiber and matrix. Because of the high complexity of vegetable fibers and their vulnerability to ecological surroundings (e.g., humidity, temperature), the assessment of numerical physical factors that could be additionally exploited in standards continues a challenge. Lastly, owing to the great variety, heterogeneity, and sophistication of raw constituents, we are presented with a multitude of systems and composite structures. Thus, the overall performance of complete bio-composites is very system-dependent, which makes the design of packaging materials with a reverse engineering approach difficult. To overcome this bottleneck, a few main parameters showing key effects of overall functional properties required to be established.

The key parameters evoked for regulating water vapor permeability (WVP) include fiber content and size, fiber/matrix adhesion, and matrix crystallinity and plasticization [59, 60]. In principle, assuming fibers are impermeable and well distributed in the matrix, WVP is likely to decrease due to tortuosity effect. In fact, lignocellulosic fibers are not impermeable. Thus, in most cases the composite WVP increased with the addition of fiber. Thus, the hygroscopic fiber character should be added to the list of the key parameters that govern WVP. Owing to aggregation and percolation phenomena WVP will increase. For small fibers and/or weak fiber/matrix adhesion the first one may occur. Weak adhesion of the fiber/matrix would also create voids in the polymers which could allow the transport of water molecules across these regions [61].

Shortly afterwards agglomeration, percolation will appear for higher fiber content and also, more likely, for higher fiber size. WVP will also increase due to a reduction in the crystallinity of the matrix and/or plasticization of the matrix caused by the addition of fiber.

Sonia and Dasan examined cellulose microfibers (CMF)/poly (ethylene-co-vinyl acetate) (EVA) with the amount of up to 13 wt.% of fiber. The introduction of low fiber content (up to 5 wt.%) resulted in an improvement in the barrier properties, while over this limit value, the 200 mm fibers induced an increase in oxygen permeability through the materials [62].

Valdés García's group studied a composite in a poly (b-caprolactone) matrix, based on almond skin residues. Since composites produced 10, 20, and 30 wt.% of 50 mm fibers, no information is given on low fiber content, and the oxygen transfer rate (OTR) showed the same evolution as Sonia and Dasan observed for highest fiber loads. In all cases, the introduction of fiber increased OTR, with a 5-times rise in the fiber content at 30 wt.%. On the contrary, Valdés García's group, measuring 10, 20, and 30 wt.% of almond skin residue content in PCL, observed a steady increase in WVP, although no major differences were noticed for smooth PCL and 10% composite PCL ($p > 0.05$). Such findings were due solely to fiber agglomeration that caused reduction in the homogeneity and cohesion of the matrix [62, 63].

Ludueña's group also reported a rise in WVP with content of cotton-based fibers (0,.5, and 15 wt.%) in PCL, but also assessed the effect of size of fibers. There is a competitive influence between the size of fibers and their water-related affinity, thus with the matrix. The smallest type of fiber (0.20 mm), being the most hydrophilic form, increases WVP composites almost to the same degree as the longest fibers (59 mm), while the medium size fiber (9.1 mm), the most hydrophobic, retains WVP composite at the same level as pure PCL [60].

2.4 Functional bio-composites for medical application

The bio-composites nowadays emanate with superior biocompatibility which perform in contemporaneous manner with the body. The architecture of these biocomposite materials is designed in such a manner that some exceptional characteristics are evolved eventually.

A combination of hydroxy apatite (HA) layer with high-density polyethylene (HDPE) as a substitution material for bone has been designed and commercialized as HAPEXTM [64–70]. In these cases, the span of HA was selected between 20 to 40 volume%. Recently, bone graft consisted of demineralized bone powder between two collagen layers was fostered and exhibited cell migration both in Vitro and in vivo investigation [71]. In another development, hydroxyapatite and a PEG/PBT (polyethylene glycol and poly-butylene terephthalate) block copolymer composites were designed with enhanced chemical linkages by using hexamethylene diisocyanate as a coupling agent. They showed that the HA particles in conjunction with polymeric matrix with covalent bonding helps in achieving bone replacement [72].

A complicated bilayer coating of graphene oxide (GO) and Poly (ε-caprolactone) (PCL)/Gelatin-forsterite nanofibers on 316 L stainless steel (SS) were developed and ultimately it showed increased suitability as orthopedic implant with improved corrosion resistance of SS [73]. However, toxicity of metallic materials is still remained as a major concern for health safety. In this connection, the biocompatibility of the scaffolds was enhanced by designing new nanocomposite system with the activation of functionalized multi-walled carbon nanotubes, kappa-carrageenan, and chitosan in hydroxyapatite (MHAp) [74].

The pursuit for targeted and coordinated drug release achieved a new dimension with the manipulation of composite structure. Nanocomposites of N-isopropyl acrylamide (NIPAAm) hydrogel with magnetic nano iron oxide particles was formulated for the pulsatile drug delivery system. By alternating the high frequency magnetic field, the heat generation in nanocomposites was controlled to regulate the swelling transition of the hydrogel [75, 76]. For another instance, nanocomposites of paclitaxel were organized using poly- (D,L-lactide-co-glycolide)/montmorillonite (PLGA/MMT) nanoparticles decorated by human epidermal growth factor receptor-2 (HER2) antibody were designed for targeted chemotherapy treatment. The NP formulation exhibited a biphasic drug release with a moderate initial burst followed by a sustained release profile. The surface decoration speeded the drug release. PLGA-MMT demonstrated increased cellular uptake by CaCo-2 and HT-29 cells [77].

With the development of biocomposite technology, there are various types of dressings for different wounds were studied for better wound healing. Consequently, many therapeutic dressings with different architecture with diverse activity have materialized and employed medically, such as natural dressings, synthetic dressings, medical dressings, and tissue engineering dressing. A promising bio-nanocomposite from nanocellulose (NC), poly(vinyl pyrrolidone) (PVP), and chitosan was fabricated by solution casting method for in vitro wound dressings [78]. The solution blended PVP and chitosan mixer formed a biocompatible combination with nanocellulose particles via hydrogen bonding. The nanocomposite showed enhanced swelling, blood compatibility and antibacterial activity. Recently, Kamel's group have fabricated distinctive biocomposite membranes from banana peel nano powder (BPnP) reinforcement in chitosan matrix. In this structure, glycerol was added as plasticizer and crosslinker to the membranes. It was found that the swelling properties of chitosan were reduced with the incorporation of BPnP. Furthermore, the results also showed that chitosan/BPnP membranes have a collaborative action with the highest activity at 10 wt% of BPnP loading [79].

3. Conclusions

In this chapter we have tried to show the usefulness of functional design concepts through variety of biocomposites design ideas and processes found in the literature. To devise a biocomposite with the expected functionality, it is indispensable to comprehend the meticulous effect of size, shape, volume fraction, orientation, distribution of filler and its compatibility with matrix. Additionally, grasping the interaction among these aspects is vital for fruitful design of biocomposites. Also cost, environmental requirements, weight, complexity and processing are the key drivers that should influence the design of biocomposites. Often, the priority of the design drivers will determine the optimum design. Consequently, this chapter has illuminated the development attained by scientists in engineering biocomposites with different functionality.

Acknowledgements

The authors would like to thank the Bangladesh University of Engineering and Technology (BUET) for providing assistance regarding for the computational facilities.

Conflict of interest

The authors declare no conflict of interest.

Author details

MD Rajbanul Akhond and Ahmed Sharif*
Bangladesh University of Engineering and Technology, Dhaka, Bangladesh

*Address all correspondence to: asharif@mme.buet.ac.bd

IntechOpen

© 2021 The Author(s). Licensee IntechOpen. This chapter is distributed under the terms of the Creative Commons Attribution License (http://creativecommons.org/licenses/by/3.0), which permits unrestricted use, distribution, and reproduction in any medium, provided the original work is properly cited.

References

[1] Pathania D, Singh D. A review on electrical properties of fiber reinforced polymer composites. :4.

[2] AL-Oqla FM, Omar AA, Fares O. Evaluating sustainable energy harvesting systems for human implantable sensors. Int J Electron. 2018 Mar 4;105(3):504-517.

[3] Salema AA, Yeow YK, Ishaque K, Ani FN, Afzal MT, Hassan A. Dielectric properties and microwave heating of oil palm biomass and biochar. Ind Crops Prod. 2013 Oct 1;50:366-374.

[4] Beneventi D, Alila S, Boufi S, Chaussy D, Nortier P. Polymerization of pyrrole on cellulose fibres using a FeCl3 impregnation- pyrrole polymerization sequence. Cellulose. 2006 Dec 1;13(6):725-734.

[5] Najar SS, Kaynak A, Foitzik RC. Conductive wool yarns by continuous vapour phase polymerization of pyrrole. Synth Met. 2007 Jan 15;157(1):1-4.

[6] Firoz Babu K, Siva Subramanian SP, Anbu Kulandainathan M. Functionalisation of fabrics with conducting polymer for tuning capacitance and fabrication of supercapacitor. Carbohydr Polym. 2013 Apr 15;94(1):487-495.

[7] Hardy JG, Lee JY, Schmidt CE. Biomimetic conducting polymer-based tissue scaffolds. Curr Opin Biotechnol. 2013 Oct 1;24(5):847-854.

[8] Balint R, Cassidy NJ, Cartmell SH. Conductive polymers: Towards a smart biomaterial for tissue engineering. Acta Biomater. 2014 Jun 1;10(6):2341-2353.

[9] Ahuja T, Mir IA, Kumar D, Rajesh. Biomolecular immobilization on conducting polymers for biosensing applications. Biomaterials. 2007 Feb 1;28(5):791-805.

[10] Cosnier S. Biomolecule immobilization on electrode surfaces by entrapment or attachment to electrochemically polymerized films. A review. Biosens Bioelectron. 1999 May 31;14(5):443-456.

[11] Cen L, Neoh KG, Li, Kang ET. Assessment of in Vitro Bioactivity of Hyaluronic Acid and Sulfated Hyaluronic Acid Functionalized Electroactive Polymer. Biomacromolecules. 2004 Nov 1;5(6):2238-2246.

[12] Guimard NK, Gomez N, Schmidt CE. Conducting polymers in biomedical engineering. Prog Polym Sci. 2007 Aug 1;32(8):876-921.

[13] Rivers TJ, Hudson TW, Schmidt CE. Synthesis of a Novel, Biodegradable Electrically Conducting Polymer for Biomedical Applications. Adv Funct Mater. 2002;12(1):33-37.

[14] Wallace GG, Smyth M, Zhao H. Conducting electroactive polymer-based biosensors. TrAC Trends Anal Chem. 1999 Apr 1;18(4):245-251.

[15] Kotwal A, Schmidt CE. Electrical stimulation alters protein adsorption and nerve cell interactions with electrically conducting biomaterials. Biomaterials. 2001 May 1;22(10):1055-1064.

[16] Lakard B, Ploux L, Anselme K, Lallemand F, Lakard S, Nardin M, et al. Effect of ultrasounds on the electrochemical synthesis of polypyrrole, application to the adhesion and growth of biological cells. Bioelectrochemistry. 2009 Jun 1;75(2):148-157.

[17] Lee JY, Bashur CA, Goldstein AS, Schmidt CE. Polypyrrole-coated electrospun PLGA nanofibers for neural tissue applications. Biomaterials. 2009 Sep 1;30(26):4325-4335.

[18] Kim D-H, Richardson-Burns SM, Hendricks JL, Sequera C, Martin DC. Effect of Immobilized Nerve Growth Factor on Conductive Polymers: Electrical Properties and Cellular Response. Adv Funct Mater. 2007 Jan 5;17(1):79-86.

[19] Garner B, Georgevich A, Hodgson AJ, Liu L, Wallace GG. Polypyrrole–heparin composites as stimulus-responsive substrates for endothelial cell growth. J Biomed Mater Res. 1999;44(2):121-129.

[20] Huang L, Zhuang X, Hu J, Lang L, Zhang P, Wang Y, et al. Synthesis of Biodegradable and Electroactive Multiblock Polylactide and Aniline Pentamer Copolymer for Tissue Engineering Applications. Biomacromolecules. 2008 Mar 1;9(3):850-858.

[21] Guiseppi-Elie A. Electroconductive hydrogels: Synthesis, characterization and biomedical applications. Biomaterials. 2010 Apr 1;31(10):2701-2716.

[22] Johnston JH, Moraes J, Borrmann T. Conducting Polymers on Paper Fibres. Synth Met. 2005 Sep 21;153(1):65-68.

[23] Johnston JH, Kelly FM, Moraes J, Borrmann T, Flynn D. Conducting polymer composites with cellulose and protein fibres. Curr Appl Phys. 2006 Jun 1;6(3):587-590.

[24] Youssef AM, El-Samahy MA, Abdel Rehim MH. Preparation of conductive paper composites based on natural cellulosic fibers for packaging applications. Carbohydr Polym. 2012 Aug 1;89(4):1027-1032.

[25] Pramila Devi DS, Bipinbal PK, Jabin T, K.N. Kutty S. Enhanced electrical conductivity of polypyrrole/polypyrrole coated short nylon fiber/natural rubber composites prepared by in situ polymerization in latex. Mater Des. 2013 Jan 1;43:337-347.

[26] Jabbour L, Chaussy D, Eyraud B, Beneventi D. Highly conductive graphite/carbon fiber/cellulose composite papers. Compos Sci Technol. 2012 Mar 8;72(5):616-623.

[27] Lively B, Bizga J, Zhong W-H. Analysis tools for fibrous nanofiller polymer composites: Macro- and nanoscale dispersion assessments correlated with mechanical and electrical composite properties. Polym Compos. 2014;35(1):10-18.

[28] Tsukada S, Nakashima H, Torimitsu K. Conductive Polymer Combined Silk Fiber Bundle for Bioelectrical Signal Recording. PLOS ONE. 2012 Apr 6;7(4):e33689.

[29] Zhan M, Wool RP, Xiao JQ. Electrical properties of chicken feather fiber reinforced epoxy composites. Compos Part Appl Sci Manuf. 2011 Mar 1;42(3):229-233.

[30] Xia Y, Yun Lu. Fabrication and properties of conductive conjugated polymers/silk fibroin composite fibers. Compos Sci Technol. 2008 May 1;68(6):1471-1479.

[31] Gelfuso MV, Silva PVG da, Thomazini D. Polypropylene matrix composites reinforced with coconut fibers. Mater Res. 2011 Sep;14(3):360-365.

[32] Jia W, Tchoudakov R, Segal E, Joseph R, Narkis M, Siegmann A. Electrically conductive composites based on epoxy resin with polyaniline-DBSA fillers. Synth Met. 2003 Jan 31;132(3):269-278.

[33] Wang W, Sain M, Cooper PA. Study of moisture absorption in natural fiber plastic composites. Compos Sci Technol. 2006 Mar 1;66(3):379-386.

[34] Ben Amor I, Rekik H, Kaddami H, Raihane M, Arous M, Kallel A. Studies

of dielectric relaxation in natural fiber–polymer composites. J Electrost. 2009 Sep 1;67(5):717-722.

[35] Amor IB, Ghallabi Z, Kaddami H, Raihane M, Arous M, Kallel A. Experimental study of relaxation process in unidirectional (epoxy/palm tree fiber) composite. J Mol Liq. 2010 Jul 31;154(2):61-68.

[36] Ben Amor I, Arous M, Kallel A. Effect of maleic anhydride on dielectric properties of natural fiber composite. J Electrost. 2014 Apr 1;72(2):156-160.

[37] Nazarzadeh Zare E, Mansour Lakouraj M, Mohseni M. Biodegradable polypyrrole/dextrin conductive nanocomposite: Synthesis, characterization, antioxidant and antibacterial activity. Synth Met. 2014 Jan 1;187:9-16.

[38] Pavlović MM, Ćosović V, Pavlović M, Talijan NM, Bojanic V. Electrical Conductivity of Lignocellulose Composites Loaded with Electrodeposited Copper Powders. Int J Electrochem Sci. 2011;6(9):3812-3829.

[39] Joussein E, Petit S, Churchman J, Theng B, Righi D, Delvaux B. Halloysite clay minerals — a review. Clay Miner. 2005 Dec;40(4):383-426.

[40] Lvov YM, Shchukin DG, Möhwald H, Price RR. Halloysite Clay Nanotubes for Controlled Release of Protective Agents. ACS Nano. 2008 May 1;2(5):814-820.

[41] Du M, Guo B, Jia D. Newly emerging applications of halloysite nanotubes: a review. Polym Int. 2010;59(5):574-582.

[42] Lvov Y, Abdullayev E. Functional polymer–clay nanotube composites with sustained release of chemical agents. Prog Polym Sci. 2013 Oct 1;38(10):1690-1719.

[43] Lvov Y, Aerov A, Fakhrullin R. Clay nanotube encapsulation for functional biocomposites. Adv Colloid Interface Sci. 2014 May;207:189-198.

[44] Abdullayev E, Lvov Y. Halloysite clay nanotubes as a ceramic "skeleton" for functional biopolymer composites with sustained drug release. J Mater Chem B. 2013 May 22;1(23):2894-2903.

[45] Abdullayev E, Abbasov V, Tursunbayeva A, Portnov V, Ibrahimov H, Mukhtarova G, et al. Self-Healing Coatings Based on Halloysite Clay Polymer Composites for Protection of Copper Alloys. ACS Appl Mater Interfaces. 2013 May 22;5(10):4464-4471.

[46] Kuroda Y, Ito K, Itabashi K, Kuroda K. One-Step Exfoliation of Kaolinites and Their Transformation into Nanoscrolls. Langmuir. 2011 Mar 1;27(5):2028-2035.

[47] Hughes AD, King MR. Use of Naturally Occurring Halloysite Nanotubes for Enhanced Capture of Flowing Cells. Langmuir. 2010 Jul 20;26(14):12155-12164.

[48] Kommireddy DS, Ichinose I, Lvov YM, Mills DK. Nanoparticle Multilayers: Surface Modification for Cell Attachment and Growth. J Biomed Nanotechnol. 2005 Sep 1;1(3):286-290.

[49] Bottino MC, Yassen GH, Platt JA, Labban N, Windsor LJ, Spolnik KJ, et al. A novel three-dimensional scaffold for regenerative endodontics: materials and biological characterizations. J Tissue Eng Regen Med. 2015;9(11):E116–E123.

[50] Liu M, Wu C, Jiao Y, Xiong S, Zhou C. Chitosan–halloysite nanotubes nanocomposite scaffolds for tissue engineering. J Mater Chem B. 2013 Mar 20;1(15):2078-2089.

[51] B. Neu HB A Voigt, R Mitlöhner, S Leporatti, CY Gao, E Donath, H

Kiesewetter, H Möhwald, HJ Meiselman. Biological cells as templates for hollow microcapsules. J Microencapsul. 2001 Jan 1;18(3):385-395.

[52] Donath E, Moya S, Neu B, Sukhorukov GB, Georgieva R, Voigt A, et al. Hollow Polymer Shells from Biological Templates: Fabrication and Potential Applications. Chem – Eur J. 2002 Dec 2;8(23):5481-5485.

[53] Kommireddy DS, Sriram SM, Lvov YM, Mills DK. Stem cell attachment to layer-by-layer assembled TiO2 nanoparticle thin films. Biomaterials. 2006 Aug 1;27(24):4296-4303.

[54] Dzamukova MR, Zamaleeva AI, Ishmuchametova DG, Osin YN, Kiyasov AP, Nurgaliev DK, et al. A Direct Technique for Magnetic Functionalization of Living Human Cells. Langmuir. 2011 Dec 6;27(23):14386-14393.

[55] Fakhrullin RF, Lvov YM. "Face-Lifting" and "Make-Up" for Microorganisms: Layer-by-Layer Polyelectrolyte Nanocoating. ACS Nano. 2012 Jun 26;6(6):4557-4564.

[56] Fakhrullin RF, Zamaleeva AI, Minullina RT, Konnova SA, Paunov VN. Cyborg cells: functionalisation of living cells with polymers and nanomaterials. Chem Soc Rev. 2012 May 15;41(11):4189-4206.

[57] Konnova SA, Sharipova IR, Demina TA, Osin YN, Yarullina DR, Ilinskaya ON, et al. Biomimetic cell-mediated three-dimensional assembly of halloysite nanotubes. Chem Commun. 2013 Apr 19;49(39):4208-4210.

[58] Pavlov AM, Geest BGD, Louage B, Lybaert L, Koker SD, Koudelka Z, et al. Magnetically Engineered Microcapsules as Intracellular Anchors for Remote Control Over Cellular Mobility. Adv Mater. 2013;25(48):6945-6950.

[59] Sanchez-Garcia MD, Gimenez E, Lagaron JM. Morphology and barrier properties of solvent cast composites of thermoplastic biopolymers and purified cellulose fibers. Carbohydr Polym. 2008 Jan 24;71(2):235-244.

[60] Ludueña L, Vázquez A, Alvarez V. Effect of lignocellulosic filler type and content on the behavior of polycaprolactone based eco-composites for packaging applications. Carbohydr Polym. 2012 Jan 4;87(1):411-421.

[61] Bessadok A, Langevin D, Gouanvé F, Chappey C, Roudesli S, Marais S. Study of water sorption on modified Agave fibres. Carbohydr Polym. 2009 Mar 2;76(1):74-85.

[62] Sonia A, Priya Dasan K. Celluloses microfibers (CMF)/poly (ethylene-co-vinyl acetate) (EVA) composites for food packaging applications: A study based on barrier and biodegradation behavior. J Food Eng. 2013 Sep 1;118(1):78-89.

[63] Valdés García A, Ramos Santonja M, Sanahuja AB, Selva M del CG. Characterization and degradation characteristics of poly(ε-caprolactone)-based composites reinforced with almond skin residues. Polym Degrad Stab. 2014 Oct 1;108:269-279.

[64] Ramakrishna S, Mayer J, Wintermantel E, Leong KW. Biomedical applications of polymer-composite materials: a review. Compos Sci Technol. 2001 Jul 1;61(9):1189-1224.

[65] Wang M. Hydroxyapatite-polyethylene composites for bone substitution: effects of ceramic particle size and morphology. Biomaterials. 1998 Dec;19(24):2357-2366.

[66] Bonfield W, Wang M, Tanner KE. Interfaces in analogue biomaterials. Acta Mater. 1998 Apr 10;46(7):2509-2518.

[67] Di Silvio L, Dalby MJ, Bonfield W. Osteoblast behaviour on HA/PE composite surfaces with different HA volumes. Biomaterials. 2002 Jan 1;23(1):101-107.

[68] Wang M, Deb S, Bonfield W. Chemically coupled hydroxyapatite-polyethylene composites: processing and characterisation. Mater Lett. 2000 Jun 1;44(2):119-124.

[69] Wang M, Bonfield W. Chemically coupled hydroxyapatite–polyethylene composites: structure and properties. Biomaterials. 2001 Jun 1;22(11):1311-1320.

[70] Ladizesky NH, Pirhonen EM, Appleyard DB, Ward IM, Bonfield W. Fibre reinforcement of ceramic/polymer composites for a major load-bearing bone substitute material. Compos Sci Technol. 1998 Mar 1;58(3):419-434.

[71] Mizuno S, Glowacki J. Three-dimensional composite of demineralized bone powder and collagen for in vitro analysis of chondroinduction of human dermal fibroblasts. Biomaterials. 1996 Sep 1;17(18):1819-1825.

[72] Ural E, Kesenci K, Fambri L, Migliaresi C, Piskin E. Poly(d,l-cactide/ε-caprolactone)/hydroxyapatite composites. Biomaterials. 2000 Nov 1;21(21):2147-2154.

[73] Khosravi F, Khorasani SN, Ghomi ER, Kichi MK, Zilouci H, Farhadian M, et al. A bilayer GO/nanofibrous biocomposite coating to enhance 316L stainless steel corrosion performance. Mater Res Express. 2019 Jun;6(8):086470.

[74] Chen M, Zhang H, Shan S, Li Y, Li X, Peng D. Fabrication of multiwalled carbon nanotubes/carrageenan-chitosan@ Ce and Sr substituted hydroxyapatite biocomposite coating on titanium: In vivo bone formation evaluations. J King Saud Univ - Sci. 2020 Jan 1;32(1):1175-1181.

[75] Patwekar SL. Nanobiocomposite: A New Approach to Drug Delivery System. Asian J Pharm AJP Free Full Text Artic Asian J Pharm [Internet]. 2016 Dec 21 [cited 2020 Dec 17];10(04). Available from: http://www.asiapharmaceutics.info/index.php/ajp/article/view/902

[76] Wen Y, Gallego MR, Nielsen LF, Jorgensen L, Everland H, Møller EH, et al. Biodegradable nanocomposite microparticles as drug delivering injectable cell scaffolds. J Controlled Release. 2011 Nov 30;156(1):11-20.

[77] Sun B, Ranganathan B, Feng S-S. Multifunctional poly(d,l-lactide-co-glycolide)/montmorillonite (PLGA/MMT) nanoparticles decorated by Trastuzumab for targeted chemotherapy of breast cancer. Biomaterials. 2008 Feb 1;29(4):475-486.

[78] Rajendran NK, Kumar SSD, Houreld NN, Abrahamse H. A review on nanoparticle based treatment for wound healing. J Drug Deliv Sci Technol. 2018 Apr 1;44:421-430.

[79] Kamel NA, Abd El-messieh SL, Saleh NM. Chitosan/banana peel powder nanocomposites for wound dressing application: Preparation and characterization. Mater Sci Eng C. 2017 Mar 1;72:543-550.

Chapter 4

Biopolymer: A Novel Bioexcipient

Sushant Kumar, Swarnima Pandey and NV Satheesh Madhav

Abstract

Polymers are the key material in design of drug delivery systems. These have been shown as the spine for drug development process. These accept an essential part in rising of novel drug delivery systems to crush different intricacies in drug delivery. These are used for controlling the appearance of the drug in needed manner. The hydrophilic and lipophilic polymers are the most ideal choice for getting the ideal conveyance in controlled, manner at the target sites. Isolated of this, these fabricated and semisynthetic polymers are made by different chemical reactions and purification measures. Since these are prepared by different unit operations which are costly. By and by days different investigates are being examined for avoiding the characteristic, physiological and reasonable issues related with the synthetic and semisynthetic polymers. So an alternative rather than synthetic and semisynthetic polymers are being investigated having interest, probability, and any leftover benefits with least troublesome ramifications for environment and physiology of the people. One of the alternatives as opposed to designed and semisynthetic polymers is biopolymers which have pulled in the thought of researchers by using an economical procedures. Biopolymers are novel, adroit and sharp polymers which have been confined from various basic sources. Biopolymers isolated from natural sources might be utilized as novel excipients having a polymeric nature. These isolated biopolymers have superb bioretardant, bio stabilizer, and mucoadhesive properties. These have the brilliant film-framing capacity and biocompatibility properties. The isolated bio-polymers have great drug release rate controlling capacities. Since these are biodegradable and might be utilized as an option in contrast to standard manufactured synthetic and semisynthetic polymers. The isolated biopolymer shows critical biodegradable, mucoadhesive, filmability, and retardability properties which are like properties of standard polymers, may be the alternative in design of novel drug delivery system design.

Keywords: Natural sources, Biomaterials, Biodegradable, Biocompatible, bioretardant, bioadhesive, biostabilizer

1. Introduction

Biocomposites are natural fiber-reinforced biopolymers. Researchers are developing these materials as an alternative to standard materials which will be nonrenewable, recalcitrant, or manufactured by pollution emitting processes. Biocomposites consist of a large variety of organic and inorganic compounds such as natural polymers, synthetic polymers, polysaccharides, sugars proteins, metals, and nanocarbon. While industrial-scale production of biocomposites is becoming more viable, the sturdiness of those natural materials limits application in many environments. Biocomposites have been described as sustainable materials because of their biodegradability, low ecotoxicity.

2. Biopolymers

Biopolymers [1] isolated from natural sources may be used as novel excipients having a polymeric nature. These isolated biopolymers have excellent bioretardant, bio stabilizer, and mucoadhesive properties. It has the excellent film-forming ability, and bio-stability properties [2]. The isolated bio-polymers have excellent drug release rate controlling abilities. Since these are natural and edible, they are biodegradable and may be used as an alternative to standard synthetic and semi-synthetic polymers [3]. The isolated biopolymer shows significant biodegradable, mucoadhesive, filmability, and retardability properties which are similar to properties of synthetic standard polymers [4]. They have most of the novel properties which can be safely used for drug delivery. The biopolymers are isolated from natural sources [5] which are economical. The synthetic polymers are prepared by using the different chemical treatment which has many harmful effects. The biopolymers have unique novel properties [6]. The biopolymers may be used for controlling the drug release in a sustained way, controlled way, extended way, prolonged way and thus are used as drug carrier bioexcipients [7]. Since they are having a natural origin and biodegradable in nature can be sued for minimizing the unwanted effects with synthetic polymers [8–10].

3. Advantages of novel biopolymer

1. Biodegradable
2. Biocompatible
3. Excellent bioretardant property
4. Biostabilizer
5. Excellent bioretardant
6. Natural
7. Economical
8. Environmental friendly
9. Excellent filmability
10. Excellent filmability

3.1 Bionanoparticles as novel nanoparticulate system

Bionanoparticles are the nanoparticles that are prepared by using the novel biocompatible and biodegradable biopolymers. We can use the novel polymeric properties in developing the bio-nano particles for targeting the drug to the brain via the blood–brain barrier in an easy way. The bio nanoparticles may release the drug to the target insignificant amount. The bio nanoparticles are stable and their excellent release rate controlling properties makes it novel [11, 12].

3.2 Novel sonication method for nanoparticles preparation

The bionanosuspension can be prepared by a novel method called the sonication method [13]. In this method, the biopolymer as bio stabilizer cum bioretardant was mixed with other ingredients like a preservative, surface active agent like PVA, nanosizent with the distilled water to make a well-dispersed suspension [14]. Then the mixture was subjected to bath sonication for 10–15 cycles to formulate the nanosized drug-loaded bionanosuspension [15].

3.3 Evaluation parameters for biopolymeric nanoformulation

A never of evaluation parameters can be performed for the prepared nanosuspension as well as bionanosuspension. The different parameters which should be considered are particle size, particle size distribution, zeta potential, particle morphology [16], dissolution study, stability study, dispersibility, % entrapment efficacy, and in vivo study [13].

The analysis of mean particle size and particle size distribution is an important parameter that defines the stability of the bionanosuspension. Nowadays the particle size and stability parameters can be evaluated by the Malvern zeta sizer. The zeta particle size gives an idea about the particle size and a particle size distribution gives an idea about the state of dispersed particle size, any agglomeration, precipitation or any lump is there.

Particle morphology and state of crystallinity is a parameter that gives an idea about for understanding any changes in drug morphology or structure on nanosizing. The amorphous drug-loaded nanoparticles can be characterized in the nanosuspension as well as bionanosuspension. This can be evaluated by x-ray powder extraction, scanning electron microscopy (SEM) characterization, transmission electron microscopy (TEM) characterization. Differential Scanning Calorimetry (DSC) is also another method for characterizing crystallinity [16].

Zeta potential measurement is another parameter for the evaluation of the particle surface charge which defines the stability of the nanosuspension as well as bionanosuspension. Zetasizer can be used for the measurement of the zeta potential. A minimum of ±30MV is generally required for the stability of nanosuspension.

The stability of nanosuspension or bionanosuspension evaluation is very important for the preparation of well-dispersed bionanosuspension. As the particle size is reduced to the nano range, the surface energy is increased and the increased surface energy may lead to the instability of nanosuspension as well as bionanosuspension.

So the uniform particle size distribution leads to the stability of nanosuspension as well as bionanosuspension.

3.4 Advantages of biopolymeric bionanoformulations over other formulations

1. Improved stability of bionanosuspension because of inbuilt properties of biopolymers.

2. Maximum drug entrapment efficacy can be obtained by designing bionanosuspension.

3. Biocompatible.

4. Biostabilizing activity.
5. Desired nanoparticles size can be prepared to cross BBB.
6. Suitable for brain targeting through ear route administration.
7. Enhanced bioavailability.
8. Drug may be released in a retardant manner.
9. Dose reduction to many folds because of longer residence.
10. Enhanced solubility of the drug.
11. Reduction of systemic toxicity.
12. Biodegradable.

3.5 Limitations of bionanoformulations

1. Stability is a big challenge for the nanosuspension for long-term storage.
2. Storage at a specified temperature.
3. Rate of sedimentation of particles during long term storage.
4. Precipitation issues during storage.
5. Accurate dose administration in form of nanosuspension.

3.6 Researches on biopolymer

Umashanker et al, NVS et al have isolated biopolymer from testa of Lallamantia royalena (Labiatae) and ready mucoadhesive biomaterial. Biopolymer was isolated by non-solvent technique. Mucoadhesivity of the biomaterial was decided by Park and Robinson and the rotating method. Spectrophotometric methods like UV, IR, TLC were conducted. The biomucoahesivity of the isolated biopolymer was confirmed by the IR method. In vitro and ex vivo evaluation was also conducted. Kala Shivani, NVS et al have formulated and evaluated bio micro dwarfs of nimesulide by isolating biomaterial from the rhizomes of (Zingiber Officinalis) common ginger. A simplified method was used for the isolation of biopolymer. IR characterization was administered for confirming the retardant activity of the biopolymer. In vitro studies, entrapment efficiency, and particle size analysis was also conducted. Tangri, Madhav [17] have isolated biomucoresident from the fruit pulp of *Artocarpus heterophyllus* commonly known as jackfruit (Moraceae), and formulated zidovudine loaded bio micro dwarfs. Particle size analysis, content uniformity, and IN-VITRO studies were conducted.IR spectrophotometry was also performed which confirmed the muco resident activity of the biomaterial. Muco retentive study was performed in the Novel Madhav Shankar study apparatus. Bisht, Upadhyay [18] have prepared polyherbal formulation for the treatment of dyslipidemia. Various sources from which polyherbal formulation was prepared are Picorrhiza kurroa, Emblica Officinalis, *Syzygium cumini*, *Trachyspermum Ammi*, Musa paradisiacal, Terminalia arjuna, pistachio, common ginger, onion, burn plant,

Eugenia caryophyllus, cereal oat. Various tests that were conducted for the evaluation of the formulation are organoleptic, Physico-chemical investigation, viscosity, surface tension, and crude fat content.

Ojha, Madhav [10] have isolated biomaterial from the fruit pulp of *Phoenix dactylifera* commonly called date palm, and evaluated its mucoadhesivity.IR spectrophotometric was performed for confirmation of the mucoadhesive nature of the biopolymer. Other tests like acute toxicity studies were performed for 14 days on rats. The shear stress method and rotating cylinder method was used for evaluating the mucoadhesive nature of the biopolymer.

Yadav, Madhv, [19] have formulated rosiglitazone bio strips for targeting trans labial drug delivery [4]. Authors have isolated biomaterial from the pulp of jackfruit (Moraceae) by the simplified economical process. Folding endurance, Thickness, content uniformity tests were conducted for evaluated various parameters of the prepared formulation. *In-vitro* drug release and stability study was conducted of the bio lip strips. Bansal Abhishek, NVS, Sharma [20] have isolated biopolymer having bio emulgent activity from the fruit pulp of Prunus institica and formulated o/w sort of emulsion. The prepared emulsion was compared with a simple emulsion having acacia as an emulsifier. Various other evaluation tests conducted are FTIR, DSC, HPLC, INVITRO drug release study. Varshney Sugandha [21] isolated biopolymer from the fruit pulp of *Manilkara zapota* and ready bio flexy films having nanosized tiagabine as a model drug. The biopolymer which was isolated from the fruit pulp of *Manilkara zapota* was used on the soft palatal surface due to its biodegradable, biocompatible, and non-irritant in nature. Spectrophotometric tests that were conducted are UV, SEM, IR, colorimetry. For the evaluation of prepared bio flex films following tests were conducted folding endurance, thickness, INVITRO drug release study.

Varshney [21] isolated a unique biopolymer from the pulp of Solano melongena and formulated bio flexy films using nanosized tiagabine as a model drug.. The biopolymer which was isolated from the fruit pulp of eggplant was used on the soft palatal surface due to its biodegradable, bio-compatible and non-irritant. Particle size analysis was through with zeta potential. Other spectrophotometric tests that were processed are UV, IR, SEM, NMR. For the prepared bio flexy films following tests were conducted folding endurance, thickness, INVITRO drug release study, weight uniformity test. Varshney [22] isolated unique biopolymer from the pulp of Ananas Cosmoses and formulated bio flexy films loaded with nanosized tiagabine as a model drug. The prepared bio flexy films were used for targeting taste bud drug delivery and therefore the biopolymer incorporated in formulation had biodegradable, biocompatible nature. Solvent casting technique was used for preparation for bio flexy films. Spectrophotometric tests that were conducted are FTIR, NMR, UV, IR, colorimetry. Other tests were conducted are muco retention time, folding endurance, weight uniformity test, thickness, swelling percentage study.

Varshney [23] isolated a unique biopolymer from the seed of pepper and ready bio flexy films to tend through taste bud drug delivery. In the prepared bio flexy films the isolated biopolymer was used as film former and therefore the bio flexy films were used for the treatment of epilepsy. Spectrophotometric tests conducted were FTIR, UV, NMR, DSC, Colorimetry. For the evaluation of prepared bio flexy films muco retention, mucoadhesion, folding endurance, thickness, content uniformity, IN VITRO drug release study.

Sugandha [24] had isolated a unique biopolymer from the petals of Rosa Polyantha and used the biopolymer to organize bio-flexy films for taste bud drug delivery. Within the bio, flexy films nano- sized Tiagabine was used as the model drug. The isolated biopolymer had inbuilt filmability, biodegradable nature, biocompatible, mucoadhesive nature, so it is often used soft palatal surface.

Tests like DSC, UV, ZETA SIZING, Colorimetry was performed. The shear stress method was used for evaluating the muco adhesivity of the biopolymer. Prepared bio flexy films were evaluated for tests like thickness, folding endurance, content uniformity, mucoadhesion, EX VIVO retention study, cell line toxicity studies. Raina and Madhav [25]. Isolated biopolymer from the berries of pepper and ready bionanosuspension using escitalopram as a drug for brain targeting through the ear. Biopolymer isolated from pepper was used as a bio retardant within the bionanosuspension. Formulations were subjected to varied tests like pH, content uniformity, release study, and EXVIVO study. Raina Deepika et al. [26, 27] had isolated biopolymer from the kernels of almond and used the isolated biopolymer for preparing bionanogel loaded with chlorpromazine for brain targeting via the nose. Prepared nanoparticles were evaluated for drug content uniformity, entrapment efficiency, IN –VITRO, muco -adhesivity, SEM, and IR. Mala et al [28] isolated biopolymer from *Cucumis sativus* (cucumber) and used the isolated bioexcipient with cefuroxime to organize bionanogel for the treatment of encephalitis [29]. The isolated biopolymer was characterized by drug-polymer interaction study, Physicochemical characterization, and acute toxicity study. The prepared bio nanoparticles were subjected to various evaluation tests like pH study, Viscosity, entrapment efficiency, IN VITRO release, in vivo release, and stability study. Kirti Singh et al [30] performed the isolation of biopolymer from the roots of *Rosa centifolia* and vinifera and prepared terbinafine loaded bioadhesive layers for the treatment of nail disease onychomycosis. Within the formulation, Beta Vulgaricus was used as a bio penetrant. The formulated films were evaluated for various parameters like nail adhesivity, folding endurance, thickness, content uniformity, and IN VITRO drug permeation and stability studies. Kirti Singh et al [31] prepared bio flexy films containing atorvastatin as model drug and biopolymer isolated from Tapioca Sago. For the evaluation of biopolymer following tests were conducted UV, IR. The prepared bio flexy films were evaluated for parameters like weight, thickness, content uniformity, folding endurance, IN VITRO drug permeation, and surface Ph. Yogita Tyagi, et al [15] performed a search work on the preparation of aripiprazole loaded bionanogel containing bio retardant isolated from the bark of *Cinnamomum verum*. Prepared bio-nano gels were targeted for the brain targeting through layers of skin meninges, transcranial nerves. For the evaluation of prepared bionanogel following tests were administered ph measurement, IN VITRO drug release, and texture, spreadability, and a couple of entrapment efficacy and stability studies. NVS, Yogita Tyagi et al [32] developed bionanogel containing nanosized aripiprazole for brain targeting. The biopolymer used as a bioretardant and bio stabilizer within the nanogel was isolated from Pudina (*Mentha arvensis*). Isolated biopolymer was evaluated spectrophotometrically by IR and UV. Following tests were performed ph measurement, surface pH study, texture, spreadability, % entrapment efficacy, and IN VITRO drug release. Singh Bhavana et al [33] conducted a search study for the preparation of bio flex films containing Venlafaxine drug for the treatment of depression. For the preparation of bio Flexi films, biopolymer was isolated from the fruits of Luffa acutangula(angled loofah). Various physicochemical tests were conducted for the evaluation of biopolymers like color, solubility, and chemical tests. Different batches of bio flexy films were evaluated for IN VITRO and in vivo drug release, folding endurance, thickness. Bioflexy films were successful in sustaining the drug release so it is often concluded that biopolymer isolated from angled loofah has promising inbuilt mucoadhesive nature. Tyagi and Madhav [34] developed bionanosuspension having biopolymer which is isolated from seeds of Buchanania lanzan (Chironzi). In the bionanosuspension, the isolated biopolymer was used as a bioemulgent. Prepared bionanosuspension was used for the treatment of depression and its safety and compatibility were proved through various evaluation tests. For the characterization of biopolymer various spectrophotometric tests like DSC, UV, IR, SEM, and NMR were used. Bionanosuspensions was evaluated with the assistance of Particle size distribution studies, IN VITRO release, pH stability studies. Tyagi and Madhav [13] developed bio nanosuspension of fluvoxamine for the management of depression. The bio nanosuspension was incorporated with the novel biopolymer *Santalum album* (sandalwood tree) and it had been designed for ophthalmic delivery of the drug. Isolated biopolymer from solvent evaporation method was subjected to varied spectrophotometric tests like IR, DSC, SEM; AND NMR. Prepared bionanosuspension was evaluated for various evaluation tests like particle size, zeta potential, entrapment efficacy, IN VITRO drug release, stability studies. Ophthalmic delivery of fluvoxamine was proved to be the novelistic approach for the treatment of depression.

Madhav [35] stated that a novel biopolymeric the material can be used to prepare drug-loaded biomicrodwarfs from Arachis hypogea seeds. The goal was to produce a product with a significant processing advantage that satisfies pharmaceutical formulators in scale-up processes. The biopolymer was isolated and characterized for its capability and efficacy to control the release of the drug. Gupta et al. [36] have reported a method for isolation of a novel biodispersant from the seeds of *Cicero arietinum* and formulation of Escitalopram granules containing bio-dispersant. Bio-dispersant was isolated by the treatment of the extract from seeds of *Cicer arietinum* with double distilled water and with ethanol and the bio-dispersant was collected and further analyzed for physicochemical properties like color, odor, particle size, shape, solubility, and IR spectral studies. The preparation of Escitalopram, granules were done using drug, lactose, bio-dispersant, bio-binder, and other processing agents. We have prepared six different formulations with varying bio-dispersant concentration and bio-binder concentrations. Tangri et al., [37] detailed a method for the formulation and evaluation of sustained-release tablets of atorvastatin by utilizing the biomaterial as a novel binder for the formulation of tablets. For the isolation of biomaterial unripe fruit pulp of *Artocarpus heterophyllus* was taken and the process of isolation used was simplified economic process. The extracted biomaterial was subjected for various physical and chemical parameters like color, color changing point, chemical tests, and I.R. spectral study. Various formulation additives were used to prepare Ibuprofen sustained-release tablets. The three atorvastatin-loaded formulations (FA1-FA3) were prepared by using different drug-polymer ratios of 1:1, 1:3, 1:5, and other excipients like starch, talc, and lactose as diluents. Erasmus et al. [38] reported that cereal grains can also be used as an agricultural raw material rich in several biopolymers. Cereal grains contain major biopolymers like starch, protein, non-starch polysaccharides, and lipids. Dry milling, wet milling, or a combination of both can be used for the primary extraction of the biopolymers. The grain is separated into its anatomical components by conventional dry milling. Anatomical components can be enriched in certain biopolymers like endosperm flour consist of approximately 80% starch. Madhav et al. [39] described the novel biomaterial from the unripe fruit pulp of *Artocarpus heterophyllus* and the evaluation of its bio-emulsifying ability by the formulation of escitalopram loaded emulsions. The isolation of biomaterial was done from the unripe fruit pulp of *Artocarpus heterophyllus* by the simple and economic process. It was subjected to various Physico-chemical parameters like color, color changing point, different chemical tests, and I.R. spectral study. Four drug-loaded emulsions were formulated (AH1-AH4) by using varying ratios of the biomaterial. Escitalopram was used as a model drug for the formulation of emulsions. Evaluation parameters like globule size, pH, the effect of centrifugation, viscosity, surface tension, creaming, freezing and thawing cycles, and *in-vitro* release were conducted on the formulated emulsions. The presence of saturated hydrocarbons, aromatic ring secondary, and tertiary alcohol groups was reported in the IR spectra of the isolated biomaterial. Singh [40] described the various components involved in pharmaceutical formulation development apart from active pharmaceutical ingredients. In recent years, the core area of research in pharmaceutical drug delivery is the excipient development because of its effect on the formulation designing development and targeted drug delivery process in various ways. Because of their low toxicity, biodegradability, stability and renewable nature biopolymers have become the choice of research as excipients. In this review, some of the most commonly used biopolymers as excipients in pharmaceutical drug delivery systems designing have been discussed. Velde et al. [41] described that to know the most suitable matrix polymer, before starting the designing it is very important to know the properties of the available polymers. It was reviewed to give information on the most suitable property of a range of biodegradable polymers. Since the data are widely scattered over many sources and are very scarce compared to the conventional polymer. Data were presented mostly as ranges as well as in graphs for quick comparison reasons. One specific application, thermoplastic pultrusion with flax as reinforcement has been also studied. Singh et al. [42] isolated the biopolymer from Tapioca sago. After isolation, it was characterized for different parameters like viscosity, ph, conductivity, and other physical characteristics. The biopolymer was also tested for the presence of carbohydrates and proteins. The isolated biopolymer was also analyzed for different spectral analysis like FTIR. The isolated biopolymer was used for the preparation of bio gel loaded with curcumin for the dermal delivery. It was concluded that the curcumin-loaded bio gel can be effectively used for the treatment of the wound by using a novel isolated biopolymer from sago as a novel retardant cum stabilizer.

3.7 Bionanoformulation in drug targeting

Madhav et al. [6] developed and evaluated duloxetine loaded bionanosuspension. The bionanosuspension was prepared by using the biopolymer isolated from *Prunus amygdalus* seeds. The biopolymer was characterized for different Physico-chemical characterization and different spectral characterization. The biopolymer was isolated by the simple economical extraction process and treatment with propanone and then solicited. The residue was recovered and dried to get the free-flowing powder. The duloxetine loaded bionanosuspension was prepared by the bath sonication method. The prepared bionanosuspension was then evaluated for different parameters like particle size, entrapment efficacy, dispersibility, zeta potential, in-vitro release, and in-vitro kinetic study, and also in-vivo study for the determination of the amount of duloxetine reached to brain via external acoustic meatus.

Madhav et al. [6] explored the feasibility of external acoustic meatus for targeting escitalopram bionanosuspension to the brain. The research reveals that escitalopram loaded bionanoparticles were found to be targeted to the brain via external acoustic meatus administration. The bionanosuspension was prepared by using the biopolymer from *Piper nigrum* in a different ratio. The biopolymer consists of the novel retardant properties to release the drug in a sustained manner. The research reveals that *Piper nigrum* can be safely used for the development of bionanosuspension for targeting the brain via external acoustic meatus. Raina et al. [6, 14, 25, 39, 43] described the preparation of duloxetine loaded bionanogel for brain targeting via external acoustic meatus. In the research work, the biopolymer was isolated from Tagetes papule and its ability in developing the duloxetine loaded bionanogel for brain targeting. In the findings of the research, the biopolymer was found to have a novelistic characteristic as polymeric nature in developing the bionanosuspension. The bionanosuspension was found to be suitable for delivering the drug to the brain. So the conclusion was that the external acoustic route can be used as the promising route for drug targeting to the brain in the treatment of depression. Madhav et al. [14] describe the formulation of chlorpromazine bionanogel by using the isolated biopolymer as a bioretardant from *Prunus amygdalus* [44]. The prepared bionanogel was evaluated for the delivery of chlorpromazine targeting to the brain. The nanoparticles were prepared by the solvent evaporation method. The formulated bionanogel were evaluated for the t50%, *in-vitro* release, *in-vivo* release study, and pharmacokinetic study. FA8 (1:15) was selected as the best formulation. Madhav et al. [6] researched the development of nanosized duloxetine via external acoustic meatus. The bionanosuspension was developed by using the biopolymer from the berries of *Piper nigrum*. The prepared bionanosuspension was evaluated for the different parameters like ph, % transmittance, content uniformity, and in vitro drug release and ex Vivo study. The obtained results were found to be significant for the treatment of CNS disorder. The drug was found to be targeted to the brain insignificant amount and this rote was found to be suitable for the delivery of duloxetine and in the treatment of depression.

Tyagi et al. [28, 45–47] researched on a new novel innovative approach for the development of the fluvoxamine-loaded bionanosuspension. The bionanosuspension was prepared by using the isolated biopolymer from the *Santalum album*. The biopolymer was characterized for *in-vitro release*, t50%, r2 values, and kinetic study to know the drug release mechanism. The results were evaluated for identifying the best fit model in drug release. Thus it was concluded that the isolated biopolymer can be suitably used for the development of stable drug-loaded bionanosuspension [16, 48–50].

3.8 Isolation and characterization of biopolymers

1. Isolation of biopolymer from natural sources.

2. Preformulation study of biopolymers system.

    a. Physical appearance, taste, shape, and texture.

    b. Solubility study.

    c. Particle size analysis by optical microscopy.

    d. Bulk density determination.

    e. Tapped density.

    f. The angle of repose.

    g. Percentage consolidation Index.

    h. Melting point testing.

    i. Chemical test.

    j. SEM analysis.

    k. FTIR spectroscopy.

l. NMR spectroscopy.

m. Mass spectroscopy.

n. DSC testing.

o. Cell line toxicity study.

3. Formulation of drug-loaded bionanoparticles as dispersed bionanosuspension using biopolymer.

3.9 Isolation of biomaterial from *the natural source*

200 gm of *natural sources like* seeds, bark, fruits, legumes, kernels, flowers are soaked overnight in purified water. The upper covering of the almond was then removed. About 100 ml of water was added to this and this mixture was mixed in a mixer. This slurry was filtered with the help of muslin cloth and thus the biomaterial was separated by filtration. The resultant was obtained as filtrate. After that, the mixture was subjected to centrifugation at about 4000 rpm for 15 minutes and then the resultant supernatant layer was properly separated and taken. Then acetone was added in the ratio of 1:1 and mixed properly. This mixture was kept in the refrigerator overnight at 4°C and the solution was centrifuged at 4000 rpm for 30 minutes. The residue was collected having biomaterial and dried in desiccators for 24 hours. This residue of biomaterial was washed with acetone and the biomaterial was dried naturally for 10 hrs for getting free-flowing powder. The collected biomaterial for stored in airtight containers after passing through sieve no. 120 for further use. The schematic flow chart of the isolation procedure has been summarized in Flow chart 4.1. This procedure was repeated six times and optimized and then the percentage yield was calculated and reported [6, 25].

4. Physicochemical characterization and evaluation of isolated biomaterial

4.1 Physicochemical characterization

The color, odor, taste of isolated biomaterial were physically evaluated. The shape of the biopolymer was also observed under the optical microscope. The color-changing point was determined by using the melting point test apparatus. The isolated biomaterial powder was filled in the capillary tube completely and it was kept in a melting point test apparatus (Cystronics). The apparatus was switch on and observed for the temperature at which color changing was observed and melting of biomaterial starts. The temperature was observed with the help of a thermometer. The organoleptic properties like color, odor, taste were observed. The pH was determined for 1% w/v aqueous solution with the help of a digital pH meter (Cystronics). The tests were performed in triplicate (n = 3) and reported [4, 51–57].

4.2 Solubility

The solubility study of the isolated biopolymers was performed in different solvents like water, acetone, methanol, ethyl acetate, 10%w/v hydrochloric acid solution, and diethyl ether and reported. The excess of the isolated biomaterial was added in 10 ml of the specific solvent system in the beaker gradually. The solution was dispersed well and kept for 24 hours on an orbital shaker for achieving an equilibrium state.

Then the solution was centrifuged at 400 rpm in the centrifuge for 10 minutes and then filtered to get the clear solution. Then the filtrate was allowed for measurement in a UV spectrophotometer machine (Mapada). The procedure was performed in triplicate for each isolated biopolymer [4, 25, 56–59].

4.3 Particle size analysis

This was performed by using the optical microscopy method. The isolated biomaterial was taken on the glass slide and added 1 drop of glycerin. The coverslip was placed on the drop and examined with the help of calibrated eyepiece micrometer under the optical microscope. During the examination, about 100 particles were counted and the particle size distribution was determined [56]. This was performed in triplicate, calculated, and reported with the help of the following equation.

$$Xg = 10\text{x}\left[\left(ni \text{ x } \log\text{Xi}\right) / \text{N}\right]. \quad (1)$$

Xg is geometrical mean diameter, ni is the number of particles in range, Xi indicates to the midpoint of the range of particle size and N refers to a total number of particles.

5. Flow property

5.1 Bulk density

The bulk density of the isolated biomaterial powder was calculated by taking accurately pre-weighed biopolymer in the measuring cylinder and then the bulk volume of the filled powder was measured. This was performed in triplicate. The bulk density was calculated and reported [56].

5.2 Tapped density

Tapped density of the isolated biomaterial was determined by taking a preweighed biopolymer in a measuring cylinder and then tapped for 100 tappings. Then the tapped volume was determined. This was performed in triplicate. The tapped density was calculated and reported [56].

5.3 Angle of repose

The angle of repose of isolated biopolymer was determined by the funnel method. Accurately weight the biopolymer was taken in the funnel. The funnel was adjusted at a height so that it just touches the apex of a heap of biopolymer powder. The powder was subjected to flow through the funnel freely on the surface. This was performed in triplicate. Thus the angle of repose was calculated and reported. The obtained results were correlated with <25- with excellent flow, 25–30 with good flow, and 30–40 –passable [60–62].

5.4 Carr's Index of compressibility

It was used for the determination of flow properties. It is a very simple, fast, and widely used method for determining powder flow characteristics. % consolidation index calculated and reported this was performed in triplicate [56].

5.5 Tapped density

According to Carr index powder with 10% flowability is considered as excellent flow characteristics. Powder with less than 15% flowability is considered as a powder with good flow characteristics [56].

6. Chemical tests of isolated biomaterial

6.1 Tests for carbohydrate

1 ml of freshly prepared biomaterial solution (5% w/v prepared biomaterial solution in double-distilled water) was taken in the test tube. Add two drops of Molisch reagent. Add 1–2 ml of conc. Sulfuric acid in the test tube and observed for the appearance of purple color at the interface of two layers formed. The test was performed and reported [4, 5, 40, 56, 63–65].

6.2 Test for protein

For testing the presence of protein in the isolated biomaterial was treated with 0.1% solution of ninhydrin reagent and 10% tannic acid solution. The presence of blue color and yellow color precipitate indicates the presence of protein. The test was performed and reported. Madhav and Yadav et al. [4] reported about the proteinous nature of biomaterials.

Biuret test was performed for the confirmation of proteins. 2 ml of *Prunus amygdalus* biomaterial was taken in the test tube (5% biopolymer solution in distilled water), add 1 ml of sodium hydroxide solution with the addition of copper sulfate solution drops. The mixture was kept aside for five minutes and observe any color changes. The appearance of violet color confirms the presence of proteins [37]. The test was performed and reported [4, 5].

6.3 Spectral analysis of the biomaterial

Spectral analysis of the isolated biomaterial was conducted like I.R, NMR, Mass spectroscopy, SEM studies. The biomaterials were subjected to IR, NMR, Mass spectroscopy studies, and the obtained spectra were interpreted and reported. SEM studies of different biomaterials were also performed and the obtained results were interpreted and reported. I.R, N.M.R, Mass spectroscopy was studied at Central Drug Research Institute, Lucknow, and SEM studies were studied at Birbal Sahani Institute of Paleobotany, Lucknow [62].

6.4 SEM (Scanning electron microscopy)

The isolated biomaterial's surface morphologies were characterized by Scanning Electron Microscope. In SEM analysis the external surface and internal structure were characterized. **Figure 1**. reveals about the flaky and rough surface of biopolymer [4].

Figure 1.
SEM of isolated biomaterial Litchi chinesis 1500x.

6.5 FTIR spectroscopy

The FTIR spectroscopy was done by preparing the KBr discs. 1 mg of isolated biomaterial was taken and mixed with 100 mg of dried and desiccated solid powder of potassium bromide. The mixture was uniformly mixed in mortar and pestle and placed in an IR lamp to remove any moisture. The mixture was converted into a disc under the pressure of 10 tons. The prepared disc was placed in the disc holder in the path of infrared radiation. In the range of 4000–200 cm^{-1}, the spectrum was recorded [4, 62] **Figure 2**. Defines the FTIR spectroscopy

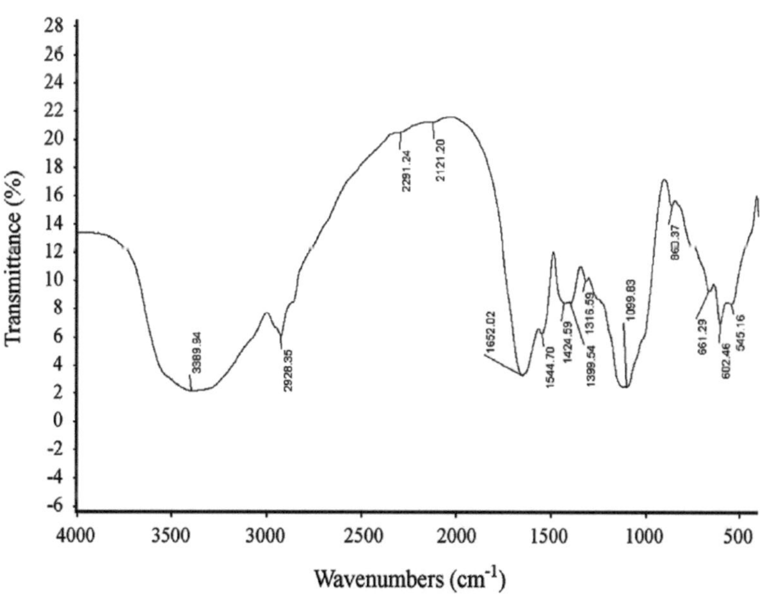

Figure 2.
IR spectrum of the isolated biomaterial Litchi chinesis.

of the isolated biopolymer from Litchi chinesis. The biopolymer showed the presence of The IR spectra revealed the presence of secondary alcohol (1099.83 cm^{-1}), aromatic nitro and phenol groups (1316.59 cm^{-1}), aromatic rings (1544.70 cm^{-1}), and the presence of alkenes (1652.02 cm^{-1}), with –CH2 and –COOH stretching (2928.35cm^{-1}) along with β-diketones and O–NO2 (1652.02 cm^{-1}). The presence of these functional groups were found to be similar with the functional groups observed in standard polymers. This functional groups revealed its polymeric nature [4].

6.6 Mass spectroscopy

It is a useful powerful technique used to quantify known materials and for identification of unknown materials and the elucidation of the structure and chemical properties of the molecule. It is an accurate method for determining the molecular mass of the compound. This is the laboratory technique in which the sample of biopolymers was introduced through the inlet system. The gas-phase ions of the compound were produced. Then molecular ion fragmentation, the ions separated in mass spectrometer according to their mass to charge ratio.

6.7 NMR spectral analysis

The NMR spectroscopy is done for spectral analysis of isolated biopolymer. The NMR spectra of different isolated biomaterials confirmed the biopolymeric nature of standard polymers [4].

6.8 Differential scanning calorimetry

In DSC testing is the thermal analysis technique in which the heat flow in or out of the sample for testing is determined as the function of temperature. Here the biopolymer sample was taken and exposed to a controlled temperature program. The glass transition temperature was determined. The heat flow range was 50–300°C. The DSC thermogram was recorded, interpreted, and reported [4, 62].

6.9 Cytotoxicity evaluation of biomaterial

Cytotoxicity evaluation of isolated biopolymer was done on the Neuroblastoma cell line. The materials used are Cell line–SHSY-5Y, (human breast cancer cell line), Ham F-12 media, Fetal Bovine Serum (FBS), the antibiotic-antimicotic solution from Thermo scientific and MTT reagent from Sigma Aldrich, USA. Tissue culture flask, 96,6 well micro-culture plates from Eppendorf, Germany. In the method, the maintenance of cell lines, the subculturing procedure of cell lines, trypsinization, cryopreservation of cell lines was done. In MTT assay the formazan product is analyzed spectrophotometrically (540 nm) after dissolution in DMSO, the spectra of treated and untreated cells giving an estimate of the extent of cytotoxicity [13, 66].

6.10 Reagents

1. 5 mg/ml MTT solution prepared in DPBS

2. Cell culture grade DMSO

6.11 Preparation of drug dilutions: (serial dilution method)

Firstly 50 mg/ml stock solution was prepared using 100% DMSO solution. From this prepared stock solution various desired final concentrations like 62.5, 125, 250, and 500 μg/ml of test compound solution was prepared as follows:

The dilution factor was 2 for the MTT assay.

For 500 μg/ml: 10 μL sol. Was taken from stock and to this 990 μl media was added.

For preparation 250 μg/ml: From 500 μg/ml solution the 500 μl was taken and was diluted with the 500 μl with media.

For preparation 125 μg/ml: From 250 μg/ml solution the 500 μl was taken and was diluted with the 500 μl with media.

For the preparation of 62.5 μg/ml: From 125 μg/ml solution the 500 μl was taken and then diluted with the 500 μl with media.

Exponentially the well-growing cell lines were collected from a 25 cm2 Tissue culture flask and a stock cell suspension of 5X104 cell/ml was prepared. A 96-well flat-bottom tissue culture plate was seeded with 5 x103 cells in 0.1 ml of F12 medium supplemented with 10% FBS and then allowed to attach for 24 hours. Test compounds were prepared just before the experiment conduction and serial dilution was done with medium to get the working stock of different 200, 100, 50, and 25 μg/ml solutions. After incubation for 24 hours, the cells were treated with 100 μl of test solutions from respective above stocks, and after treating the cells were incubated for 48 hrs. The cells present in the control group received only the medium containing the 0.5–0.25% DMSO. Each treatment procedure was performed in triplicates. After the treatment duration, 30 μL5 mg/ml MTT solution was then added to each well, and the whole was incubated for 3 hours at 37°C in maintained sterile conditions. After the completion of incubation time, the MTT containing media was removed carefully from all wells then formazan crystals were dissolved by adding 100 μL of DMSO. The plate was shaken for 5 minutes on a gyratory shaker machine and the optical density was noted at 540 nm in an ELISA plate reader. The percentage of cell viability was calculated. O.D of each well was read and expressed as % cell death: (Absorbance of control wells- absorbance of test wells/absorbance of control wells) x 100. Results were expressed as the mean ± S.E.M. The O.D values (proportional to cell death) were plotted against the tested drug concentrations and then interpreted [3, 4, 13, 34, 45].

7. Formulation of drug loaded bio-nanoparticulate system

The different formulations of bio-nanosuspension were prepared by using different drug-biopolymer ratio and drug-standard polymer ratio. The bionanosuspension was prepared by sonication of the mixture of drug and biopolymer along with other excipients like polyvinyl alcohol as suspending agent, sodium benzoate as the preservative, purified water, and dextrose as nanosizent [9, 13, 67, 68]. The lamotrigine, biopolymer, and other excipients were accurately weighed and triturated with the addition of the double-distilled water. This mixture was sonicated for 3 cycles. Then 0.5 ml of 0.5% polyvinyl alcohol was added during sonication. The volume of the formulations was made up to 10 ml with double distilled water having sodium benzoate 0.1–0.5%). Add dextrose if necessary as a nanosizing agent and allowed for sonication for 15 cycles at 4000 rpm. After sonication, the bionanosuspension was refrigerated for two days [11, 41, 69, 70]. If no settlement is there then it means the formulation is optimized. If the settlement is there, then 0.5 ml of 0.5% polyvinyl alcohol was again added and allowed for sonication for 10 cycles and refrigerated for 48 hours. The different formulations were prepared and after optimization, according to stability, the formulations were prepared. In the same way, the nanosuspension was also prepared by using lamotrigine with a standard polymer like hydroxypropyl methylcellulose [3, 4, 13].

8. Future aspects

1. The isolated biopolymers still have not explored for their novel inbuilt characteristics in drug delivery, which can be used as an alternative to standard polymers as these are biodegradable, biocompatible, and bio-retardant cum bio stabilizer in nature [71–75].

2. The biopolymers may be isolated in economical ways from the different edible natural sources [76–78].

3. Thus economically isolated biopolymer can be safely used for developing the most suitable and stable drug-loaded bio-nanosuspension in drug targeting in a very significant way route with maximum patient compliances [36, 38, 57, 79–81].

4. Thus the isolated biopolymers have a number of novel properties which can be used as a novel bioexcipient in design of novel drug delivery design. Apart of these inbuilt novel properties these isolated biopolymer are economical in production. The current researches and finding provide an alternative to standard polymers and retardant excipients in very economical rate of production with a number of smart inbuilt bioretardant cum biostabilizing properties.

Author details

Sushant Kumar[1*], Swarnima Pandey[2] and NV Satheesh Madhav[3]

1 Faculty of Pharmacy, Pharmacy college Saifai, Uttar Pradesh University of Medical Sciences, Etawah, U.P., India

2 Goel Institute of Pharmacy and Sciences, Affiliated to Dr. A.P.J. Abdul Kalam Technical University, Lucknow, U.P., India

3 Formulation Research Development, Vital Therapeutics and Formulations, Pvt. Ltd., Telangana, Hyderabad, India

*Address all correspondence to: k.sushant25@gmail.com

IntechOpen

© 2021 The Author(s). Licensee IntechOpen. This chapter is distributed under the terms of the Creative Commons Attribution License (http://creativecommons.org/licenses/by/3.0), which permits unrestricted use, distribution, and reproduction in any medium, provided the original work is properly cited.

References

[1] L. Goswami, N.V.S. Madhav, K., Upadhyaya, Development and evaluation of bionanaoparticles as novel drug carriers for delivery of selegiline. *Internation Current Pharmaceutical Journal*, 5(4), (2016) 33-37.

[2] Jorda, J., Marechal,P., Rigal, L. and Pontalier, P.Y., 2002. Biopolymer purification by ultrafiltration. *Desalination*, 148, pp. 187-191.

[3] Tyagi, D. and Madhav, N.V.S., 2019. Smart innovative approacheds for the designing fluvoxamine loaded bionanosuspension for the management of depression. *International Journal of Applied Pharmaceutics,* 11(1), pp. 191-197.

[4] Madhav, N. V. S. and Yadav, A.P., 2013. A novel translabial platform utilizing bioexcipients from *Litchi chinesis* for the delivery of rosiglitazone maleate. *Acta Pharm Sin B,* 3, pp. 408-415.

[5] Kokate, C.K.,1991. *Practical Pharmacognosy*, 3rd ed. New Delhi, India, Vallabh Prakashan.

[6] Madhav, N.V.S. , Jaiswal, V. and Ojha, A., 2017. Development and evaluation of nanosized aripiprazole-loaded bioflexy films using a biopolymer from *Lagenaria siceraria* for brain delivery through orosoft palatal mucosal platform, *Egypt Pharmaceut J,* 16 (1) pp. 62-68.

[7] Desai , S., Perkinsa, J., Harrisonc, B.S. and Sankar, J., 2010. Understanding release kinetics of biopolymer drug delivery microcapsules for biomedical applications. *Materials Science and Engineering B*, 168, pp. 127-131.

[8] Iwona, M. and Maria, M., 2012. The release of active substances from selected carbohydrate biopolymer Membranes. *Carbohydrate Polymers* 87, pp. 2432– 2438.

[9] Vashista, A., Guptab, Y.K. and Ahmad, S., 2012. Interpenetrating biopolymer network based hydrogels for an effective drug delivery system. *Carbohydrate Polymers*, 87, pp. 1433– 143.

[10] Ojha A. and Madhav, N.V.S (2012) Isolation and characterization of novel mucoadhesive biomaterial from Phoenix dactylifera, International Current Pharmaceutical Journal 2012, 1(8): 205-208

[11] Yoncheva, K., Lizarraga, E. and Irache, J.M., 2005. Pegylated nanoparticles based on poly (methyl vinyl ether-co-maleic anhydride): preparation and evaluation of their bioadhesive properties. Eur J Pharm Sci, 24, pp. 411-9.

[12] Elzoghby, A.O., Abd-Elwakil, M.M., Abd-Elsalam, K., Elsayed, M.T., Hashem, Y. and Mohamed, O., 2016. Natural Polymeric Nanoparticles for Brain-Targeting: Implications on Drug and Gene Delivery. Curr Pharm Des, 22(22), pp. 3305-3323.

[13] Tyagi, Y., and Madhav, N. V. S., 2020. Inbuilt novel bioretardant feature of biopolymer isolated from cucumis sativa for designing drug loaded bionanosuspension. *Journal of Drug Assessment,* 9(1), pp.72-81.

[14] Madhav, N.V.S., and Raina, D., 2016. Formulation of chlorpromazine bio-nanogel using a biretardant from *Prunus amygdalus* for nose to brain targeting. *International Journal of Advanced Research in Biological Sciencs*. 3(2), pp. 151-157.

[15] Tyagi, D. and Madhav, N.V.S., 2019 Formulation and evaluation of nanosized aripiprazole-loaded bionanogels using novel bio-retardant from bark of Cinnamomum verum. J. Presl. Curr Med Drug Res, 2019, 1 (1), Article ID 172

[16] Muller, R.H. and Grau, M.J., 1998. Increase of dissolution rate and solubility of poorly water soluble drugs

as nanosuspension. Proceedings. World Meeting APGI/APV, Paris, 2, pp. 62-624.

[17] Tangri P. and Madhav, N. V. S., 2011, Formulation and evalution of Zidovudine biomicrodwarfs using a Novel biomucoresident from Artocarpus heterophyllus. International Journal of PharmTech Research, Vol. 3, No.1, pp. 169-174

[18] Bisht Upadhyay, NVS et al. 2011. Phytochemical screening and standardization of polyherbal formulation for dyslipidemia, International Journal of Pharmacy and Pharmaceutical Sciences, 3, 235-238

[19] Yadav Pratap Abhishek, NVS 2013 A novel translabial platform utilizing bioexcipients from Litchi Cinesis for the delivery of rosiglitiazione maleate, Acta Pharmaceutica Sinica B, Volume 3, Issue 6, 408-415

[20] Bansal Abhishek, NVS, Sharma AK, (2014) Isolation and Characterization of Biopolymer Obtained from Fruit Pulp of Prunus insititia, Scholars Research Library, Der Pharmacia Lettre, 2012, 4 (3):927-930

[21] Varshney Sugandha, NVS (2017) Design and Evaluation of Bio-Flexy Films using a Novel Biopolymer from Manilkara zapota Loaded with Nanosized Tiagabine Journal of Moleculer Medicine and Clinical Application, Volume: 2.1,1-5

[22] Varshney Sugandha, NVS (2017) Development and Evaluation of Unidirectional Mucoadhesive Bio-Flexy Films Loaded with Nanosized Topiramate using a Novel Biopolymer from Glycine max, Indian Journal of Pharmaceutical Education and Research |Vol 54 | Issue 3 | Jul-Sep, 2020

[23] Varshney, S. and Madhav, N.V. S., 2018. Development and evaluation of Bioflexy films using a novel biopolymer from Ananas cosmosus loaded with nanosized tiagabine. Egyptian Pharmaceutical Journal, 17 (1), pp. 1-12.

[24] Varshney Sugandha, NVS 2018 Development and Evaluation of Mucoadhesive Bio-Flexy Films Using a Novel Biopolymer from Solanum melongena Loaded with Nanosized Topiramate, Journal of Nanomedicine & nr uoJ Biotherapeutic Discovery, 8:1Volume 8 • Issue 1 • 1000158

[25] Raina D. and Madhav, N.V.S., 2017. A smart approach for formulation process of escitalopram loaded bio-nano suspension for brain delivery via novelistic EAM (External acoustic meatus) platform. *Indian Drugs,* 54(08), pp. 81-84.

[26] Raina Deepika and NVS 2018 formulation and evaluationof duloxetine loaded bio nano gels ffor effective delivery to the brain via external acoustic meatus, J.Bio.Innov 7 (6), pp: 843-857, 2018

[27] Raina Deepika and NVS 2016 formulation and performance evaluation of escitalopram loaded bionano suspension using a novel bio-retardent from piper nigrum International journal of life sciences and Review Vol. 3(5): 60-66.

[28] Madhav, N.V.S., Raina, D. and Mala, M., 2015. Design of cefuroxime loaded bionano gels for brain specificity via ear. Journal of Advanced Drug Delivery, 2(4), pp. 1-5.

[29] Madhav, N.V.S. and Raina, D., 2017. Formulation and evaluation of Duloxetine loaded bio-nano suspension for brain specificity via acoustic meatus. *SOJ Pharm Pharm Sci*., 4(1), pp. 1-5.

[30] Singh Kirti, NVS 2017 Smart ungual biopenetrant from the roots of Beta vulgarisournal of Applied Pharmaceutical Research. 2017, 5(2): 21-26

[31] Singh Kirti , NVS 2018 An Innovative Gateway to Deliver

Nanosized Atorvastatin by BioFlexy Film Approach Global Journal of Medical Research: B Pharma, Drug Discovery, Toxicology & Medicine Volume 18 Issue 5 Version 1.0 Year 2018

[32] Yogita Tyagi & NV Satheesh Madhav (2020) Inbuilt novel bioretardant feature of biopolymer isolated from cucumis sativa for designing drug loaded bionanosuspension, Journal of Drug Assessment, 9:1, 72-81.

[33] Singh Bhawana, Kumar Pranay, NVS (2017) Formulation and evaluation of venlafaxine-loaded bio-flexi film for brain specificity via oro-trans soft palatal route, Curr Med Drug Res, 2017, 1 (1), Article ID 174

[34] Tyagi, Y. and Madhav, N. V. S., 2019. Smart innovative approach for designing fluvoxamine loaded bio-nanosuspension for the management of depression. *International Journal of Applied Pharmaceutics*, 11(1), pp. 191-197.

[35] N.V. Satheesh Madhav and M. S. Uma Shankar. A Novel biopolymeric smart stabilizer from Cocos Nucifera meat kernel and its pharm

[36] Gupta, R. and Madhav, N.V.S., 2011. Formulation and Evaluation of Escitalopram Solid dispersion using Biopolymer from *Cicer arietinum* Seeds, *Int.J. PharmTech Res,* 3(3), pp. 1227-1233.

[37] Tangri, P. and Madhav, N.V.S.,2012. Formulation & Evaluation of Atorvastatin Loaded ER Tablets, *Der Pharmacia Lettre,* 4 (3), pp. 833-839.

[38] Erasmus, C., and Taylor, J. R. N., 2003. Large-scale extraction of cereal biopolymers. In: Afripro, Workshop on the Proteins of Sorghum and Millets: Enhancing Nutritional and Functional Properties for Africa. Paper 16. Conference proceedings published online at afripro.org.uk.

[39] Madhav, N.V.S., and Tangri, P., 2011. Formulation and evaluation of zidovudine bio-micro dwarfs using a novel bio-muco resident from *Artocarpus heterophyllus*. *Int J Pharm Tech Res*., 3(1), pp. 169-174.

[40] Singh, A.V., 2011. Biopolymers in Drug Delivery: A Review. *Pharmacologyonline,* 1, pp. 666-674.

[41] Velde, K.V. and Kiekens P., 2002. Biopolymers: overview of several properties and consequences on their applications. *Polymer Testing,* 21, pp. 433-442.

[42] Singh, S. and Madhav, N.V.S., 2019. Design and evaluation of curcumin smart biogels loaded with Tapoica sago biopolymer for novel dermal delivery. World Journal of Pharmaceutical Research, 8, 11, pp. 1187-1201.

[43] Madhav, N.V.S., and Raina, D., 2017. An innovative approach for delivery of nanosized dulexitine via external acoustic meatus (EAM) platform. Gazi medical Journal, 28(4), pp. 1-5.

[44] Madhv, N.V.S., and Raina, D., 2017. Formulation and performance evaluation of escitalopram loaded bionano-suspension using a novel bioretardant from *Piper nigrum*. *Interational Journal of Life Sciences and Review,* 3(5), pp. 60-66.

[45] Tyagi, Y. and Madhav, N. V. S., 2019. Design of selegiline-loaded bio-nanosuspension for the management of depression using novel bio-retardant from *Manilkara zapota*, Drug Development and Industrial Pharmacy, 45:8, pp. 1351-1360.

[46] Tyagi, Y. and Madhav, N. V. S., 2019. Design of selegiline loaded bionano-suspension for the management of depression using novel bio-retardant from Manilkara zapota, Drug Development and Industrial Pharmacy, 45: 8, pp. 1351-1360.

[47] Tyagi, D. and Madhav, N.V.S., 2019. Smart innovative approaches for the

designing fluvoxamine loaded bionano-suspension for the management of depression. International journal of applied pharmaceutics, 11(1), pp. 191-197.

[48] Masihuddin, Jafri, M. A., Siddiqui, A. and Khan, A., 2018. Phytochemistry, Pharmacological Activities And Traditional Uses of *Prunus Amygdalus* With Special Reference Of Unani Medicine: An Updated Review. *International Journal of Scientific Research and Review*, 7(11), pp. 83-92.

[49] Muthukumaran, S., Tranchant, C., Shi, J., Ye, X. and Xue, S. J., 2017. Ellagic acid in strawberry (Fragaria spp.): Biological, technological, stability, and human health aspects. *Food Quality and Safety,* 1(4), pp. 227-252.

[50] Nag, M., Mukherjee, S, Roy, D., Paul, R. K. and Sahidu, M., 2018. Extraction and Chemical tests on *Cicer Arietinum* seed collected from North Bengal Region of West Bengal, *India. PharmaTutor,* 6(10), pp. 1-4.

[51] Oha, J. K., Leea, D, I. and Park, J.M., 2009. Biopolymer-based microgels/ nanogels for drug delivery applications. *Progress in Polymer Science*, 34, pp. 1261-1282.

[52] Özen, F., Temeltaş, H., Aksoy, Ö., 2012. The anatomy and morphology of the medicinal plant, *Lilium candidum* l. (Liliaceae), distributed in marmara region of turkey. *Pak. J. Bot.*, 44(4), pp. 1185-1192.

[53] Parmar, K., Singh, N. K. and Tinde, L. K., 2016. Farming of kidney beans (Rajma) and its health benefits. *Rashtriya Krishi,* 11 (2), pp. 1-2.

[54] Patocka, J. and Navratilova, Z., 2019. Bioactivity of *Lilium candidum* L : A Mini Review. *Biomed J Sci & Tech Res,* 18(5), pp. 13859-13861.

[55] Pavan, G., Sravanthi, K. and Reddy, D. R., 2015. Evaluation of hepatoprotective activity of *Lilium candidum*. in animals. *Int J Trends in Pharm & Life Sci*, 1(4)., pp. 516-534.

[56] Sinko, P.J., 2006. *Martin's, Physical pharmacy and pharmaceutical sciences*, 5th Edition, B.I publications Pvt. Ltd, New Delhi.

[57] *Indian Pharmacopeia*, 1996. Ministry of Health and family welfare, Government of India, Controller of Publications, New Delhi.

[58] Raina D, Madhav NVS. Formulation and evaluation of bionanogel for effcetive lelivery to brain via external acoustic meatus. Journal of bio. Innov. 2018;7(6): 843-857.

[59] Raut, R. A. and Ghotankar, A. M., 2019. Review of Khuskhus (khaskhas) (seeds of *Papaver somniferum* linn.) with special reference of ayurveda medicine. *Journal of Ayurveda and Integrated Medical Sciences*, 4(6), pp. 228-231.

[60] Shah, T. I., Sharma, E. and Ahmad, G., 2014. *Juglans regia* Linn: A Phyto-pharmacological Review. *World Journal of Pharmaceutical Sciences*, 2(4), pp. 364-373.

[61] Shaista, A. and Amrita, P., 2016. Delicate, fragrant, lady of the night- A medicinal gift. Journal of Medicinal Plants Studies, 4(6), pp. 13-17.

[62] Silverstein, R.M., Francis, X., *Webster.*, David, J., Kiemle. and David, L. Bryce., 2010. *Spectroscopic identification of organic compounds,* 6th ed., New Delhi; Wiley India Pvt. Ltd.

[63] Singh, S. and Madhav, N.V.S., 2019. Design and evaluation of curcumin smart biogels loaded with Tapoica sago biopolymer for novel dermal delivery. *World Journal of Pharmaceutical Research*, 8, 11, pp. 1187-1201.

[64] Singh, S., Verma, V., Yadav, R. and Singh, B., 2018. Pharmacognostical

study of Amalaki (Emblica officinalis Gaertn.). *Journal of Pharmacognosy and Phytochemistry*, 7(3), pp. 3476-3480.

[65] Soares, A. G.,2014. Cultivation, Antioxidant Properties and Health Benefits, Edition: 1th, Chapter: Postharvest Technologies for Fresh and Minimally Processed Strawberry Conservation, *Publisher: Nova Science Publishers, Inc., New York, Editors: Nathan Malone,* pp.189 – 308.

[66] Kumar, S., N. V. Satheesh Madhav, A. Verma, and K. Pathak. "Development and Evaluation of Novel Bionanoparticles Loaded With Nanosized Lamotrigine by Using the Novel Isolated Biopolymer from Phaseolus Vulgaris Seeds". International Journal of Pharmaceutical Sciences and Drug Research, Vol. 12, no. 3, May 2020, pp. 247-54.

[67] Varshney, S. and Madhav, N.V. S., 2018. Development and evaluation of Bio-Flexy films using a novel biopolymer from Ananas cosmosus loaded with nanosized tiagabine. *Egyptian Pharmaceutical Journal*, 17 (1), pp. 1-12.

[68] Varshney, S. and Madhav, N.V. S., 2018. Development and Evaluation of Mucoadhesive Bio-Flexy Films Using a Novel Biopolymer from *Solanum melongena* Loaded with Nanosized Topiramate. *J Nanomedine Biotherapeutic Discov*., 8: (1), pp.1-10.

[69] Vijay, K. P., Laxman, B.C., Balasaheb, S. R., Yuvraj, N. R. and Janardhan, P. M., 2013. Pharmacognostic, physicochemical and phytochemical investigation of *Tagetes erecta* linn flowers (Asteraceae). *Journal of Biological & Scientific Opinion*, 1 (1), pp. 21-24.

[70] Zainish, M. S., Salman, A. and Muhammad, M. H., 2016. *Phaseolus vulgaris* linn.: botany, medicinal uses, phytochemistry and pharmacology. *World Journal of Pharmaceutical Research*, 5, (11), pp. 1611-1616.

[71] Zhang, J., Xie, Z., Zhang, N. and Zhong, J., 2017. Nanosuspension drug delivery system: preparation, characterization, postproduction processing, dosage form, and application. *Nanostructures for Drug Delivery*, pp.413-443.

[72] Abdullah, Khalid, M. and Hussainm, M. K., 2017. Badam (*Prunus amygdalus* Bail.): A Fruit with Medicinal Properties. *International Journal of Herbal Medicine,* 5(5), pp.114-117.

[73] Al-Reza, S. M., Rahman, A., Cho, Y. and Kang, S. C., 2010. Chemical Composition and Antioxidant Activity of essential oil and organic extracts of *Cestrum nocturnum*. *Journal of Essential Oil and Bearing Plants.* 2010; 13 (5) :615 – 624.

[74] Al-Snafi, A. E., 2016. The medical Importance of *Cicer arietinum* - A review. *IOSR Journal Of Pharmacy,*6(3), pp. 29-40.

[75] Arora, M., Singh, S. and Kaur, P., 2013. Pharmacognostic & Phytochemical Evaluation Of Selected Seeds Of '*Cicer arietinum*' Linn. Seeds From Roopnagar Punab. *International Journal of Pharmaceutical Science Invention*, 2 (11), pp. 18-29.

[76] Basha, S. K., Mohammad, M., Shifa, M.D., Srilekha, Aisha, S.K. and Raja, R. R., 2016. Poppy seed-An Overview. *RA Journal of Applied Research,* 2(10), pp. 670-672.

[77] Esfahlan, A. J., Ostadrahimi, A. and Tabibiazar, M. A., 2019. Comparative Review on the Extraction, Antioxidant Content and Antioxidant Potential of Different Parts of Walnut (*Juglans regia* L.) Fruit and Tree. Molecules, 24, pp. 2133.

[78] Fatima, T., Showkat, U. and Hussain, S. J., 2018. Nutritional and health benefits of walnuts. *Journal of Pharmacognosy and Phytochemistry,* 7(2), pp. 1269-1271.

[79] Gopi, G., Elumalai, A. and Jayasri, P., 2012. Aconcise review on *Tagetes erecta*. *International Journal of Phytopharmacy Research*, 3(1), pp. 16-19.

[80] Goswami, L., Madhav, N.V.S. and Upadhyaya, K., 2016. Development and evaluation of bio-nanoparticles as novel drug carriers for the delivery of Selegiline. International *Current Pharmaceutical Journal*, 5(4), pp. 33-37.

[81] Hasan, R., Islam, N. and Islam, R., 2016. Phytochemistry, pharmacological activities and traditional uses of Emblica officinalis: A review. *International Current Pharmaceutical Journal,* 5(2), pp. 14-21.

Chapter 5

Properties of High-Density Polyethylene-Polypropylene Wood Composites

Mourad Saddem, Ahmed Koubaa and Bernard Riedl

Abstract

We investigated the effects of polymer blend variation on the physical, mechanical, and thermal properties of wood-polymer composites (WPC). We used high-density polyethylene (HDPE) and polypropylene (PP) and a combination of 80% PP, 20% HDPE, and 80% HDPE, 20% PP as polymer blends for WPC formulations to simulate recycled plastics. We used black spruce (*Picea mariana* Mill.) hammer milled fibers (75–250 μm) at 35 wt% as a filler for all the formulations. A two-step process was used for WPC manufacturing; pellet extrusion followed by test samples injection. Tensile and three bending tests characterized the WPC mechanical properties. Thermogravimetric analysis (TGA) and differential scanning calorimetry (DSC) characterized the WPCs' thermal properties. Water absorption and contact angle measurements assessed the composite dimensional stability. Infrared spectroscopy (FTIR) and electron scanning microscopy (SEM) investigated the WPCs' surface chemistry and microstructure. Mechanical properties and dimensional stability varied according to polymer composition, with better performance for WPC containing higher PP proportions. Thermal properties varied with the polymer composition in the WPC, with better thermal stability for the formulation containing higher HDPE proportions. Surface chemistry analysis did not reveal any chemical changes on the WPCs surface. Scanning electron microscopy analysis revealed distinct phases in all WPCs without evidence of interfacial adhesion.

Keywords: wood fiber, polypropylene, high-density polyethylene, wood-polymer composites, mechanical properties, water absorption, surface chemistry, microstructure

1. Introduction

Significant progress was achieved during the last years in wood-polymer composites (WPC) research and product development. In addition to the environmental advantages, the use of wood fiber in WPC manufacturing has several benefits, including low cost, renewability, biodegradability, low specific gravity, and high specific strength and stiffness [1, 2]. WPC use in construction and transport is in constant evolution due to these advantages [1–3]. The increase in environmental awareness increased the demand for environmentally friendly products [4]. WPC presents an excellent alternative to plastic products with lower environmental impact and cost [5]. Lately, many studies have tried to improve WPC properties and competitiveness by introducing new additives and processes [6–9]. However,

IntechOpen

the polymers' prices are very volatile and depend on petroleum prices, affecting the competitiveness of such products. Thus, the economic aspect is also crucial to ensure the sustainability of WPC production. Using recycled materials appears to be an excellent way to lower the raw material cost and consequently the total production cost. The recycled thermoplastic polymer is commonly used as a matrix to produce WPC [10]. Moreover, the environmental advantages of using recycled polymers could decrease the WPC production costs.

In landfills, plastics are not biodegradable. The 2019 global production of plastics was 368 million tons. This important production leads to a large plastic waste stream, making recycling a great challenge [11, 12]. Recycling plastics in relatively long lifecycle products such as WPC could decrease the carbon footprint of the non-renewable polyolefin and avoid their burial in landfills [13]. Waste prevention comes to be an excellent way to avoid recycling energy uses and facilities expenses. In some cases, waste prevention could be less favorable by diminishing the downstream material recycling and preventing the low-impact of secondary production [14]. In recycling centers, sorting wastes is the most expensive part of the recycling process. Recycled plastics are always a mix of different kinds of polymers, such as polyethylene terephthalate (PET), low-density polyethylene (LDPE), high-density polyethylene (HDPE), and polypropylene (PP). These polyolefins have different chemical compositions making them incompatible during processing which could significantly affect the properties of the resulting products.

Several reports [4, 15–21] investigated the effects of recycling on the properties of WPC, both components of these composites could be from recycled materials. Ashori and Sheshmani [15] showed that recycled newspapers in recycled PP composites showed maximum water absorption compared to non-recycled components formulation. Low et al. [4] used epoxy to produce recycled cellulose fiber epoxy composites. Results showed an improvement in mechanical properties. Tajvidi and Takemura [17] showed that recycled fibers increase the composites hydrophobicity due to better fiber-polymer adhesion. Nerenz et al. [16] showed that the addition of a sunflower hull diminished the tensile strength of composites compared to neat PP. Xiaolin et al. [18] studied the feasibility of composites made with recycled newspapers and magazines and recycled PP. The study showed that recycled fibers and PP could be a viable source for producing WPC.

Petroleum-based polymers such as PE, HDPE, PET, and PP serve for WPC production. These petroleum-based polymers are also large-scale products used for many applications such as packaging and constitute the main component of waste landfills.

Many studies investigated the potential of the polyolefin recycled matrix components for WPC [10, 21–36]. Chtourou et al. [21] showed that municipal waste plastics composed of 95% PE and 5% PP and pulp fibers produced WPC with average tensile properties compared to virgin polyolefin. Li et al. [22] used PET and PE to produce microfiber-reinforced composites. They reported that the mechanical properties of the PET composites greatly improved compared to the typical PET/PE blend at the same composition. Cui et al. [23] showed good compatibility of the treated wood post-consumer fiber with recycled HDPE and additives, giving WPC good mechanical properties. Beg and Pickering [24] studied the effect of eight times reprocessing on wood PP composites. The study showed a decrease in mechanical properties with the increase in reprocessing, and the thermal stability increased with the repeated process. In another study, Beg and Pickering [25] found that the equilibrium of moisture content of WPC decreases after eight times reprocessing. Ares et al. [26] showed that with more than 10% wood flour content, the reprocessed PP composites showed mechanical and rheological properties similar to those of virgin polymer composites. González-Sánchez et al. [27] showed that the fiber dispersion is not dependent on the polyolefin type, and reprocessed PP showed more pseudo-plasticity loss than the reprocessed PP HDPE. Bhaskar et al. [28] used recycled PP to produce WPC and compared the mechanical properties with virgin PP WPC. Recycled PP with less than 50% fiber load had WPC with good properties [29]. Combining shell core with recycled polyolefin could lead to cost-effective advantageous WPC [30]. Catto et al. [31] showed that recycled polyolefin is a viable alternative due to its comparable physical and mechanical properties to virgin polyolefin. Adding virgin PP to recycled HDPE improves WPC properties [32]. De Oliveira Santos et al. [33] showed that using recycled PET below the melting temperature could enhance the composites processing and mechanical properties. Another study [34] reported that virgin and recycled polymers give similar mechanical properties and water uptake.

The reuse of polyolefins also has environmental advantages. It contributes to decreasing global warming and using non-renewable fossil hydrocarbons [10]. However, the emissions of reprocessed products weaken these advantages. Rangavar et al. [35] reported that recycled PP leads to WPC with similar or even better properties than the virgin polymer in addition to economic and environmental benefits. Thus, recycled polymers exhibit a high potential to replace virgin polymers for producing fiber-filled thermoplastic composites [36].

This study assessed the effect of the variation of PP and PEHD proportions on WPC properties. We used two different combinations of these two polymers (20% HDPE + 80% PP and 80% HDPE + 20% PP) to simulate recycled plastics. We did not use any grafting agent in all formulations to better assess their interaction and the effects of their simultaneous presence on the WPC properties.

2. Experimental

Black spruce fibers were hammer milled with a Wiley Laboratory Mill mounted with a 2 mm opening sieve. Fibers were classified with Ro-tap Laboratory Sieve Shaker to obtain fibers with size in the class of 200–60 mesh (75–250 μm) and oven-dried to reach 3% moisture content. Polymers used in this study are high-density polyethylene (HDPE) (DOW DMDA-8907 NT7, Dow Chemical) and polypropylene (PP 4150H, Pinnacle polymers, USA). The HDPE is a semi-crystalline material (typically 70–80%) has a 0.95 density, 9.0 g/10 min melt index, and a 135°C melting point. The PP has a density of 0.90 g/cm^3 and a melt flow index of 55 g/10 min at 230°C.

Composite pellets were processed by a counter-rotating, intermeshing, conical twin-screw extruder (Thermo Scientific HAAKE PolyLab OS Rheodrive 7 with Rheomex OS extruding module). The screw speed was 50 rpm, and the barrel and die temperature was 165°C. The wood fibers proportion in the composites was maintained constant at 35 wt %. Composites were cooled in a water bath and cut up in pellets of 3 mm size. Injection molding with Minijet Haake injection machine produced test samples for bending and tensile tests using a mold temperature of 80°C, and barrel and nozzle temperature of 175°C.

Three-point bending properties were measured on 1 mm thick, 10 mm wide, and 60 mm long test samples and at a test speed of 1.4 mm/min according to the ASTM D 790 standard. Tensile properties were measured on dog-borne shaped samples of 2 mm thickness, 4 mm width, and 75 mm length at a 5 mm/min speed according to the ASTM D 638.

WPC water uptake was assessed according to ASTM 1037 on bending type specimens in triplicates. Samples were soaked in distilled water, and their weight was measured periodically up to 45 days immersion using a laboratory balance with an accuracy of ±0.001 g.

A Shimadzu IR Tracer-100 (Kyoto, Japan) served for FTIR spectroscopy. The analyzed spectrum was 400–4000 cm^{-1} with a resolution of 1 cm^{-1}. We used the software lab solutions IR de Shimadzu with 50 scans for each measurement. Thermogravimetric analysis (TGA) and differential scanning calorimetry (DSC) of the WPC samples were conducted in NETZCH leading thermal analyzer using a heating rate of 10°C/min under helium with 25 mL/min flow rate from room temperature to 900°C.

Scanning electron microscopy served to analyze fractured surfaces of WPC tensile specimens. Micrographs were generated at 20 keV at a 5 mm working distance with Hitachi S3500 (Tokyo-Japan) electron microscope. Samples were prepared and gold-coated before observations.

The mechanical properties were subjected to an analysis of variance using the ANOVA procedure of the IBM SPSS statistics software. Polymer variability was the studied factor. Effects and differences between means were considered statistically significant at $p < 0.05$. ANOVA assumptions were verified using graphical diagnostics and Levene test for equal variances.

3. Results and discussions

3.1 Mechanical properties

Table 1 presents the tensile and bending properties for the studied polymers and composites. Polymer variability significantly affected all WPC mechanical properties (**Table 2**). PP showed better mechanical properties compared to the HDPE. Pure HDPE samples showed the lowest tensile and flexural properties (**Table 1**) compared to PP and WPC. Adding wood fibers to HDPE, PP or polymer mixes improved the tensile and flexural moduli of elasticity and strengths and decreased the elongation at maximum strength (**Table 1**). Several factors explain this decrease, including the stiff nature of the wood fiber, the poor adhesion between the fiber and the polymers, and the incompatibility of the non-miscible polymeric chains in the case of WPC made with a mix of polymers. The heterogeneity of composition, poor adhesion and lack of polymers' miscibility lead to increased microstructure cavities and voids, which negatively affect the WPC strength and ductility.

3.2 Physical properties

Water absorption increases with the duration of immersion and remains constant upon saturation [37]. **Figure 1** illustrates the water absorption for the investigated formulations after 45 days of immersion in distilled water. HDPE and PP samples maintained a constant weight. They did not absorb water after 45 days of immersion (**Figure 1**) due to the polymers' hydrophobicity, weak surface energy, and free hydroxyl groups' absence.

For WPCs, water uptake increased with time of immersion according to the same pattern of evolution. PP-based composites showed lower water uptake compared to HDPE-based composites. The water uptake of polymers mix-based composites is in between.

Wood fibers are responsible for water absorption in WPCs because of their hydrophilic character. Adding wood fiber to PP, HDPE, or polymer mixes increased the water uptake in agreement with previous findings [37–41]. The WPC water absorption phenomenon is due to the capillary transport into the gaps and flaws at the interfaces between the fibers and polymers because of poor fiber-polymer adhesion, incomplete wettability, and impregnation, which lead to water transport by micro-cracks formed during the processing [39–41].

Properties	Formulations					
	HDPE WPC	HDPE-PP WPC		PP WPC	HDPE	PP
		80-20	20-80			
Tensile properties						
E_t, GPa	1.11 ± 0.05	1.58 ± 0.09	1.77 ± 0.12	1.85 ± 0.21	1.02 ± 0.12	1.27 ± 0.16
σ_t, MPa	24.7 ± 0.4	23.6 ± 0.2	28.28 ± 0.98	30.21 ± 0.83	20.41 ± 0.88	31.38 ± 1.65
ε, %	3.18 ± 0.07	4.19 ± 0.37	3.60 ± 0.46	3.86 ± 0.14	6.39 ± 0.42	7.79 ± 0.94
Flexural properties						
E_f, GPa	0.72 ± 0.06	0.86 ± 0.05	1.40 ± 0.03	1.52 ± 0.05	0.40 ± 0.02	1.12 ± 0.04
MOR, MPa	26.1 ± 1.1	28.8 ± 0.5	38.7 ± 1.0	42.3 ± 0.4	17.2 ± 0.6	40.1 ± 1.5

E_t: tensile modulus of elasticity, σ_t: tensile strength, ε: elongation at maximum strength, E_f: flexural modulus of elasticity, MOR: modulus of rupture in flexion.

Table 1.
Average and standard deviation of HDPE, PP, and WPCs' tensile and flexural properties.

Properties	F value
Tensile modulus	94.4**
Tensile strength	40.3**
Elongation at maximum yield	36.3**
Flexural modulus	431.0**
Flexural strength	543.3**

***Significant at 1% probability level.*

Table 2.
Analysis of variance (F values) on the effect of polymer variability on WPC properties.

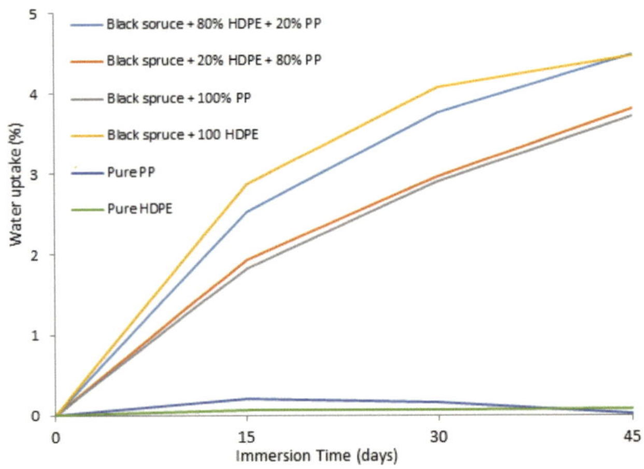

Figure 1.
Evolution of water uptake of WPC made from black spruce fibers and HDPE, PP polymers and their mixes after 45 days of water immersion.

3.3 Thermal stability

Figure 2 illustrates the TGA curves of all studied formulations. The maximum degradation of PP occurred at 490°C, while the maximum degradation of the WPC made with 100% PP occurred at 520°C. The same tendency occurs for the WPC made with HDPE. These results indicate that the presence of the wood fibers improves the thermal stability of PP and HDPE. For all studied formulations, total degradation occurs at around 600°C. The lowest degradation is obtained for the composites made with 20% PP and 80% HDPE polymer mix. This composite was the most thermally stable among the different composites. The HDPE and PP curves show a one-stage degradation (**Figure 2**) and WPC's two stages of degradation (**Figure 2**). The first stage corresponds to the wood fiber component, which begins to degrade at 220°C, and the second stage corresponds to the polymer degradation. The obtained patterns of variation of the TGA curves are typical of those reported for WPC [42].

Table 3 shows the DSC results for the tested polymers and WPCs, including the melting temperature (T_f), the enthalpy of fusion (ΔH_f), and the crystallinity index (X_c), the crystallization temperature (T_c), and the enthalpy of crystallization (ΔH_c). WPCs made with polymers mixed showed two different fusion peaks. The melting temperatures of 80% HDPE + 20% PP and 20% HDPE + 80% PP WPCs were 132.3°C and 165.7°C, respectively. The HDPE WPC showed the highest

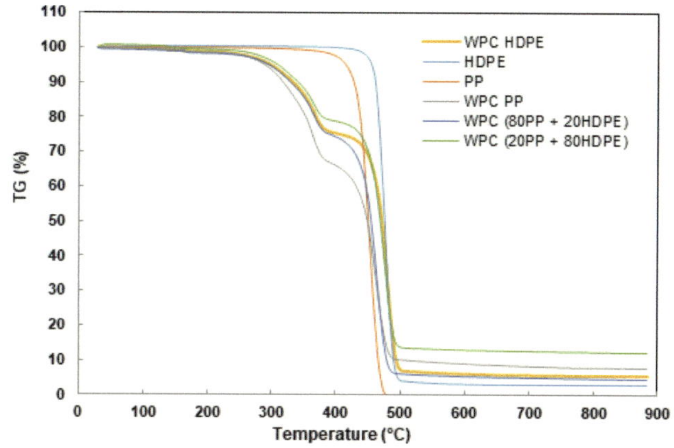

Figure 2.
TGA curves of HDPE and their WPCs made with black spruce fiber and HDPE, PP, (80% HDPE + 20% PP) and (20% HDPE + 80% PP) matrices.

Samples	T_f (°C)	ΔH_f (J/g)	X_c (%)	T_c (°C)	ΔH_c (J/g)
Polymers					
HDPE	136.4	157.8	54.4	114.2	159.5
PP	167.7	118.2	56.1	110.1	123.1
WPC 35% black spruce fibers					
HDPE	130.8	75.8	74.5	116.4	87.8
PP	165.2	46.9	64.1	115.68	59.5
80% HDPE-20% PP	132.3	73.8	72.7	115.43	81.1
20% HDPE + 80%	165.7	51.3	69.7	115.60	62.8

T_f: the melting temperature, ΔH_f: the enthalpy of fusion, X_c: the crystallinity index, T_c: the crystallization temperature, ΔH_c: the enthalpy of crystallization.

Table 3.
Thermal properties of pure HDPE, pure PP, and WPC made with black spruce fiber and HDPE, PP, (80% HDPE + 20% PP) and (20% HDPE + 80% PP) matrices.

crystallinity index (X_c = 74.5%) due to its better thermal stability than PP WPC (X_c = 64.1%). The 20% HDPE-80% PP showed lower crystallinity (X_c = 69.7%) than the 80% HDPE-20% PP WPC PP (X_c = 72.7%). Thus, increasing PP proportion in the polymer mix decreases the crystallinity.

Pure PP showed the lowest crystallization temperature at 110.1°C (**Table 3**). The crystallization temperature of all WPCs is higher than the pure polymers crystallization temperature because of the degradation of the wood fibers during the heating process. Adding wood fibers decreased the fusion and the crystallization enthalpies due to the dilution effect of the wood fiber within the polymers. The decrease in the polymer content reduces the heat of fusion, and the increase in wood fiber content limits the thermal movement of the polymer molecular chain and results in a reduction in released heat fusion is reduced.

3.4 Surface chemistry

Figure 3 shows the FTIR absorbance spectra range (4000–400 cm^{-1}) of wood, HDPE, PP, and the studied WPC formulations. Spectra of the wood fibers are similar to those previously reported [43, 44]. These spectra show the presence of a

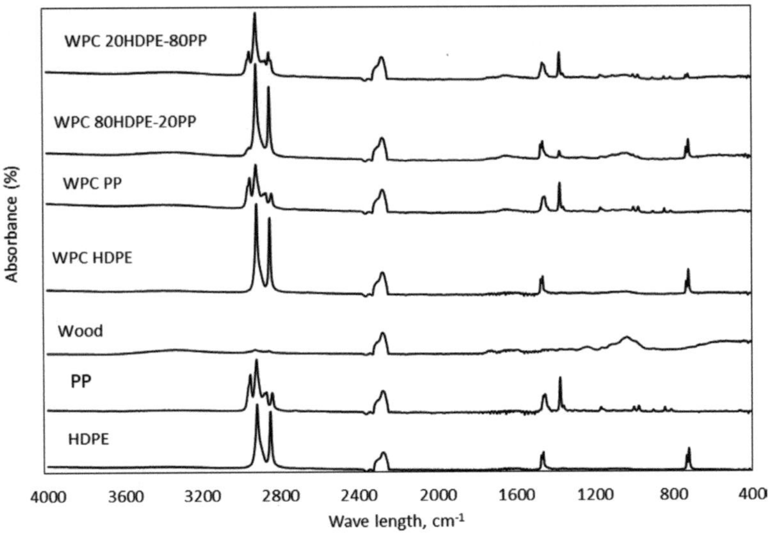

Figure 3.
FTIR spectra of wood; HDPE, PP, and WPCs made with HDPE, PP, and PP-HDPE mixtures.

broad stretching band for intermolecular bonded hydroxyl groups (OH) at around 3400 cm^{-1}. The OH groups may include absorbed water, aliphatic primary and secondary alcohols found in carbohydrates and lignin, aromatic primary and secondary alcohols in lignin and extractives, and carboxylic acids in extractives [43]. This OH stretching band is flanked by prominent methylene/methyl bands appearing at around 2900 cm^{-1}. These bands are shifted and divided into two peaks at 2922 cm^{-1} and 2853 cm^{-1}, respectively. An ester carbonyl vibration occurs at about 1728 cm^{-1}, emanating from carbonyl (C=O) stretching of acetyl groups in hemicelluloses and carbonyl aldehyde in lignin and extractives. This vibration emanates from the carbonyl (C=O) stretching of carboxyl groups in hemicelluloses, lignin, and extractives, as well as esters in lignin and extractives [43]. Between 1500 and 400 cm^{-1}, we observe several absorption bands due to various functional groups of wood constituents. The bands around 1457 cm^{-1}, 1424 cm^{-1}, and 1373 cm^{-1} are associated with methylene deformation and methyl asymmetric and methyl symmetrical vibrations [43]. The strong bands appearing at 1270 cm^{-1} are due to either a carbon single-bonded oxygen stretching vibration or an interaction vibration between carbon single-bonded oxygen stretching and in-plane carbon single-bonded hydroxyl bend in carboxylic acids [43].

Papp et al. [44] attributed bands containing no other nearby absorption maxima to one chemical component (1510 cm^{-1}: aromatic rings, 1270 cm^{-1}: guaiacyl units, 1158 cm^{-1}: C–O–C bonds of cellulose). The absorption band at 1158 cm^{-1} is due to the asymmetric stretching of C–O–C in the cellulose and hemicelluloses [43] or the saturated fatty acid ester carbon single-bonded oxygen stretching associated with the ester carbonyl at lower wavenumber [43]. The strong intensity bands at 1059 cm^{-1} and 1036 cm^{-1} correspond to cellulose [43]. The vibrations between 896 cm^{-1} and 810 cm^{-1} are due to ring stretching and out-of-plane carbon single-bonded hydrogen [43].

The infrared spectrum of HDPE shows peaks at 2916 cm^{-1} and 2849 cm^{-1} associated with methylene asymmetric and symmetric C–H stretching, respectively. The peak at 1472 cm^{-1} is due to the methylene asymmetrical C–H bending, while the peak at 1463 cm^{-1} is associated with methylene scissoring. The peaks at 730 and 720 cm^{-1} are associated with crystalline and amorphous methylene. The absorption peaks on the infrared spectrum of PP are related to the methyl group ($-CH_3$) and the methylene group. The typical peak at 1375 cm^{-1} is associated with the symmetric bending vibration mode of methyl group CH_3. The Peak located at 840 cm^{-1} is assigned to C−CH_3 stretching vibration, while the absorption peaks displayed at 972, 997, and 1165 cm^{-1} are associated with −CH_3 rocking vibration. The peak observed at 2952 cm^{-1} is related to −CH_3 asymmetric stretching vibration. The absorption peaks at 1455, 2838, and 2917 cm^{-1} are attributed to −CH_2− symmetric bending, −CH_2− symmetric stretching, and −CH_2− asymmetric stretching, respectively [45–48].

The HDPE WPC is similar to that of HDPE, with only minor changes in the spectrum. This similarity is because the polymer coats the fiber. The PP WPC is also identical to that of PP for the same reason. For the WPC made with polymer mixes, the 80% HDPE-20% PP WPC FTIR spectrum is similar to the HDPE WPC, while

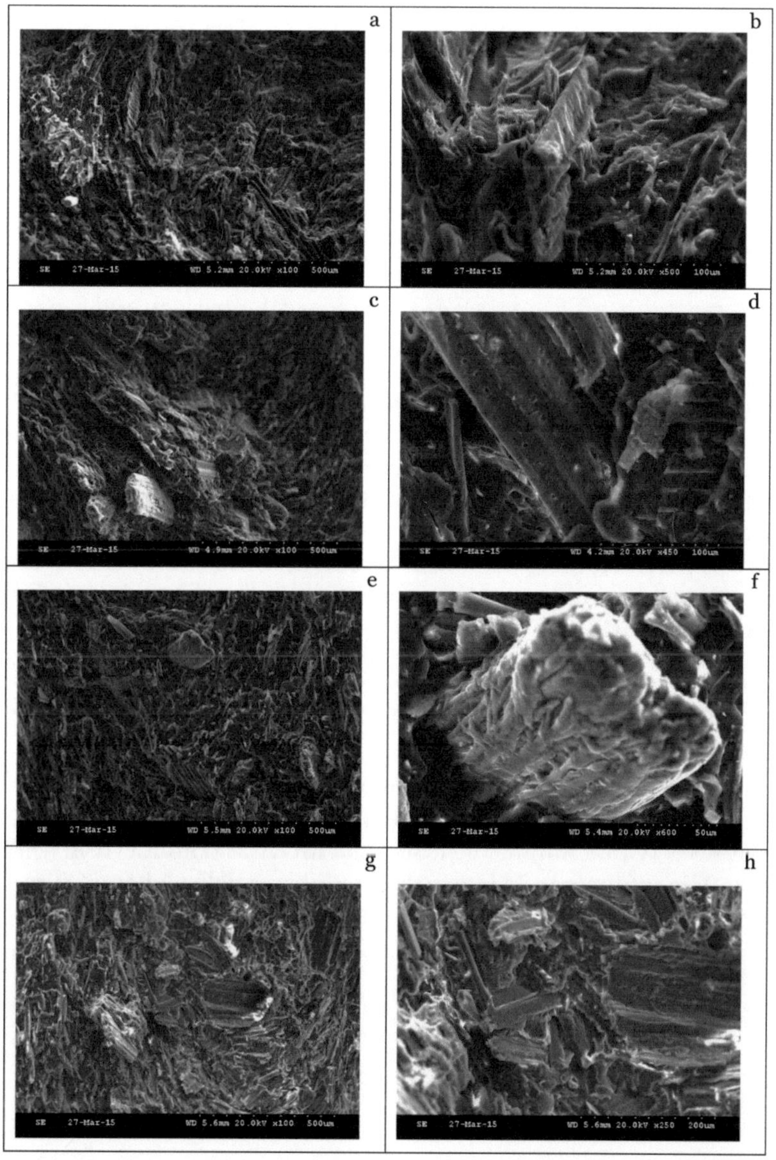

Figure 4.
Scanning Electron microscopic observation of fractured surfaces of the different WPCs made with black spruce fiber (35 wt%) and HDPE ((a) ×100; (b) ×500); PP ((c) ×100; (d) ×450); 80% PP-20% HDPE ((e) ×100; (f) ×600); 20% PP-80% HDPE ((g) ×100; (h) ×250).

20% HDPE-80% FTIR spectrum is similar to that of PP. The disappearance of peaks associated with wood in the two spectra is also due to the fibers coating with the polymer mixes.

Among the slight differences between the composites and the polymer, the spectrum is the absorbance peak at 1031 cm^{-1}. This peak is associated with the carbon-oxygen (C–O) bonding between cellulose and hemicellulose.

3.5 WPC microstructure by scanning electron microscopy

SEM observations (**Figure 4**) show distinct phases without evidence of interfacial adhesion. Voids and traces of pullout appear in all figures indicating the weak interfacial adhesion between the different phases. **Figure 4d** shows a wood fiber (the element with punctuations) and a crack between this element and the polymer, demonstrating poor contact and adhesion in the interface. **Figure 4e, f** and **h** shows a complete fiber pulled out from the polymers, confirming the weak interfacial adhesion. In addition, the bad dispersion seen in **Figure 4e–g** confirms the non-compatibility of the two polymers and the absence of interfacial adhesion between the polymers and the wood fibers.

4. Conclusions

Polymer variability affected the WPC mechanical, physical, and thermal properties. The WPC formulations with higher PP proportions exhibited higher mechanical properties and dimensional stability than those with higher HDPE proportions. Surface chemical and microstructure analysis showed the lack of adhesion between the different phases. The polymers coated the wood fibers without chemical reaction, and PP and HDPE were not miscible.

The polymer mix simulated recycled plastic. Although WPC made with this mix showed lower mechanical properties and dimensional stability than the composites made with one polymer, these composites from these mixes are suitable for several end-uses, sustainable, and have the economic advantage of being made from recycled polymers.

Acknowledgements

Canada Research Chairs Program (Grant 557752), Natural Sciences and Engineering Research Council of Canada (Grant 567663), ForvalueNet NSERC Strategic Network (Grant number 504736), and MITACS (Grant IT03894) funded this research. The authors would like to thank Williams Belhadef and Gilles Villeneuve for their technical assistance and Sebastien Migneault for scientific support.

Conflict of interest

The authors declare no conflict of interest.

Author details

Mourad Saddem[1], Ahmed Koubaa[1*] and Bernard Riedl[2]

1 University of Quebec in Abitibi-Temiscaming, Rouyn-Noranda, Quebec, Canada

2 Laval University, Quebec City, Quebec, Canada

*Address all correspondence to: ahmed.koubaa@uqat.ca

IntechOpen

© 2021 The Author(s). Licensee IntechOpen. This chapter is distributed under the terms of the Creative Commons Attribution License (http://creativecommons.org/licenses/by/3.0), which permits unrestricted use, distribution, and reproduction in any medium, provided the original work is properly cited.

References

[1] Ashori A. Wood-plastic composites as promising green-composites for automotive industries! Bioresource Technology. 2008;**99**(11):4661-4667

[2] Mu B, Wang H, Hao X, Wang Q. Morphology, mechanical properties and dimensional stability of biomass particles/high-density polyethylene composites: Effect of species and composition. Polymers. 2018;**10**(3):308

[3] Migneault S, Koubaa A, Perré P. Effect of fiber origin, proportion, and chemical composition on the mechanical and physical properties of wood-plastic composites. Journal of Wood Chemistry and Technology. 2014;**34**(4):241-261

[4] Low IM, Somers J, Kho H, Davies I, Latella B. Fabrication and properties of recycled cellulose fiber-reinforced epoxy composites. Composite Interfaces. 2009;**16**(7-9):659-669

[5] Teuber L, Osburg VS, Toporowski W, Militz H, Krause A. Wood polymer composites and their contribution to cascading utilization. Journal of Cleaner Production. 2016;**110**:9-15

[6] Pérez E, Famá L, Pardo S, Abad M, Bernal C. Tensile and fracture behaviour of PP/wood flour composites. Composites Part B: Engineering. 2012;**43**(7):2795-2800

[7] Poletto M, Zeni M, Zattera AJ. Effects of wood flour addition and coupling agent content on mechanical properties of recycled polystyrene/wood flour composites. Journal of Thermoplastic Composite Materials. 2012;**25**(7):821-833

[8] Rodríguez-Llamazares S, Zuñiga A, Castaño J, Radovic LR. Comparative study of maleated polypropylene as a coupling agent for recycled low-density polyethylene/wood flour composites. Journal of Applied Polymer Science. 2011;**122**(3):1731-1741

[9] Zhang J, Koubaa A, Xing D, et al. High-performance lignocellulose/polycarbonate biocomposites fabricated by in situ reaction: Structure and properties. Composites Part A Applied Science and Manufacturing. 2020;**138**:106068

[10] Vantsi O, Karki T. Environmental assessment of recycled mineral wool and polypropylene utilized in wood polymer composites. Resources Conservation and Recycling. 2015;**104**(Part A):38-48

[11] Gao H, Xie Y, Ou R, Wang Q. Grafting effects of polypropylene/polyethylene blends with maleic anhydride on the properties of the resulting wood-plastic composites. Composites Part A: Applied Science and Manufacturing. 2012;**43**(1):150-157

[12] Gu F, Guo J, Zhang W, Summers PA, Hall P. From waste plastics to industrial raw materials: A life cycle assessment of mechanical plastic recycling practice based on a real-world case study. Science of the Total Environment. 2017;**601-602**:1192-1207

[13] Valente M, Sarasini F, Marra F, Tirillò J, Pulci G. Hybrid recycled glass fiber/wood flour thermoplastic composites: Manufacturing and mechanical characterization. Composites Part A: Applied Science and Manufacturing. 2011;**42**(6):649-657

[14] Cleary J. A life cycle assessment of residential waste management and prevention. International Journal of Life Cycle Assessment. 2014;**19**(9):1607-1622

[15] Ashori A, Sheshmani S. Hybrid composites made from recycled materials: Moisture absorption and

thickness swelling behavior. Bioresource Technology. 2010;**101**(12):4717-4720

[16] Nerenz BA, Fuqua MA, Chevali VS, Ulven CA. Processing and characterization of a polypropylene biocomposite compounded with maleated and acrylated compatibilizers. International Journal of Polymer Science. 2012;**2012**:472078. DOI: 10.1155/2012/472078

[17] Tajvidi M, Takemura A. Effects of reprocessing on the hygroscopic behavior of natural fiber high-density polyethylene composites. Journal of Applied Polymer Science. 2011;**122**(2):1258-1267

[18] Xiaolin Z, Xiangfeng B, Rumin W. Study on mechanical properties and water absorption behaviour of wastepaper fibre/recycled polypropylene composites. Polymers and Polymer Composites. 2013;**21**(6):395-402

[19] Shao Y, Moras S, Ulkem N, Kubes G. Wood fibre-cement composites by extrusion. Canadian Journal of Civil Engineering. 2000;**27**(3):543-552

[20] Bootklad M, Chantarak S, Kaewtatip K. Novel biocomposites based on wheat gluten and rubber wood sawdust. Journal of Applied Polymer Science. 2016;**133**(30):1-6. DOI: 10.1002/APP.43705

[21] Chtourou H, Riedl B, Ait-Kadi A. Reinforcement of recycled polyolefins with wood fibers. Journal of Reinforced Plastics and Composites. 1992;**11**(4):372-394

[22] Li Z-M, Yang M-B, Lu A, Feng J-M, Huang R. Tensile properties of poly (ethylene terephthalate) and polyethylene in-situ microfiber reinforced composite formed via slit die extrusion and hot-stretching. Materials Letters. 2002;**56**(5):756-762

[23] Cui Y, Lee S, Noruziaan B, Cheung M, Tao J. Fabrication and interfacial modification of wood/recycled plastic composite materials. Composites Part A: Applied Science and Manufacturing. 2008;**39**(4):655-661

[24] Beg MDH, Pickering KL. Reprocessing of wood fibre reinforced polypropylene composites. Part I: Effects on physical and mechanical properties. Composites Part A: Applied Science and Manufacturing. 2008;**39**(7):1091-1100

[25] Beg MDH, Pickering KL. Fiber pretreatment and its effects on wood fiber reinforced polypropylene composites. Materials and Manufacturing Processes. 2006;**21**(3):303-307

[26] Ares A, Bouza R, Pardo SG, Abad MJ, Barral L. Rheological, mechanical and thermal behaviour of wood polymer composites based on recycled polypropylene. Journal of Polymers and the Environment. 2010;**18**(3):318-325

[27] González-Sánchez C, Fonseca-Valero C, Ochoa-Mendoza A, Garriga-Meco A, Rodríguez-Hurtado E. Rheological behavior of original and recycled cellulose–polyolefin composite materials. Composites Part A: Applied Science and Manufacturing. 2011;**42**(9):1075-1083

[28] Bhaskar J, Haq S, Yadaw SB. Evaluation and testing of mechanical properties of wood plastic composite. Journal of Thermoplastic Composite Materials. 2012;**25**(4):391-401

[29] Yang TH, Leu SY, Yang TH, Lo SF. Optimized material composition to improve the physical and mechanical properties of extruded wood-plastic composites (WPCs). Construction and Building Materials. 2012;**29**:120-127

[30] Kim BJ, Yao F, Han G, Wang Q, Wu Q. Mechanical and physical properties of core-shell structured wood-plastic composites: Effect of shells

with hybrid mineral and wood fillers. Composites Part B: Engineering. 2013;**45**(1):1040-1048

[31] Catto AL, Stefani BV, Ribeiro VF, Santana RMC. Influence of coupling agent incompatibility of post-consumer HDPE in thermoplastic composites reinforced with eucalyptus fiber. Materials Research. 2014;**17**(Suppl. 1): 203-209

[32] Arango C, RodrÍGuez-Llamazares S, CastaÑO J, Zuniga A. Effect of virgin heterophasic PP copolymer content on moisture absorption and thermal and mechanical properties of recycled polyethylene wood flour composites. Journal of the Chilean Chemical Society. 2014;**59**(1):2374-2377

[33] De Oliveira Santos RP, Castro DO, Ruvolo-Filho AC, Frollini E. Processing and thermal properties of composites based on recycled PET, sisal fibers, and renewable plasticizers. Journal of Applied Polymer Science. 2014;**131**(12): 1-13 DOI: 10.1002/APP.40386

[34] Sommerhuber PF, Welling J, Krause A. Substitution potentials of recycled HDPE and wood particles from post-consumer packaging waste in wood-plastic composites. Waste Management. 2015;**46**:76-85

[35] Rangavar H, Taghiyari HR, Ghofrani M, Khojaste-Khosro S. Improving physical and mechanical properties in particleboard by recycled polyethylene and canola residues. International journal of Environmental Science and Technology. 2016;**13**(3): 857-864

[36] Bajwa DS, Bajwa SG, Holt G, Srinivasan R, Coffelt T, Nakayama F, et al. Recycling of ligno-cellulosic and polyethylene wastes from agricultural operations in thermoplastic composites. Waste and Biomass Valorization. 2014;**5**(4):709-714

[37] Kord B. Investigation on the long-term water absorption behavior and cell morphology of foamed wood-plastic nanocomposites. Journal of Thermoplastic Composite Materials. 2012;**27**(3):379-394. DOI: 10.1177/0892705712446800

[38] Borysiak S. Influence of wood mercerization on the crystallization of polypropylene in wood/PP composites. Journal of Thermal Analysis and Calorimetry. 2012;**109**(2):595-603

[39] Célino A, Fréour S, Jacquemin F, Casari P. The hygroscopic behavior of plant fibers: A review. Frontiers in Chemistry. 2014;**1**:1-12

[40] Karmaker AC. Effect of water absorption on dimensional stability and impact energy of jute fibre reinforced polypropylene. Journal of Materials Science Letters. 1997;**16**:462-464. DOI: 10.1023/a:1018508209022

[41] Dhakal HN, Zhang ZY, Richardson MOW. Effect of water absorption on the mechanical properties of hemp fibre reinforced unsaturated polyester composites. Composites Science and Technology. 2006;**67**:1674-1683

[42] Bouafif H, Koubaa A, Perré P, Cloutier A, Riedl B. Wood particle/high-density polyethylene composites: Thermal sensitivity and nucleating ability of wood particles. Journal of Applied Polymer Science. 2009;**113**(1): 593-600

[43] Bouafif H, Koubaa A, Perré P, Cloutier A, Riedl B. Analysis of among-species variability in wood fiber surface using DRIFTS and XPS: Effects on esterification efficiency. Journal of Wood Chemistry and Technology. 2008; **28**(4):296-315

[44] Papp G, Barta E, Preklet E, Tolvaj L, Berkesi O, Nagy T, et al. Changes in DRIFT spectra of wood irradiated by UV laser as a function of

energy. Journal of Photochemistry and Photobiology A: Chemistry. 2005; **173**(2):137-142

[45] Larkin P. Infrared and Raman Spectroscopy: Principles and Spectral Interpretation. Amsterdam: Elsevier; 2017

[46] Andreassen E. Infrared and Raman spectroscopy of polypropylene. In: Karger-Kocsis J, editor. Polypropylene: An A-Z Reference (Part of the Polymer Science and Technology Series, POLS, Volume 2). Berlin: Springer; 1999. pp. 320-328. DOI: 10.1007/978-94-011-4421-6

[47] Luongo J. Infrared study of polypropylene. Journal of Applied Polymer Science. 1960;**3**(9):302-309

[48] Sakai K, Sobue H. Study of structure and thermal properties of polypropylene and chlorinated polypropylene by infrared spectroscopy and differential scanning calorimetry. Journal of Applied Polymer Science. 1972;**16**(10):2657-2670

Chapter 6

Characterization, Modeling and the Production Processes of Biopolymers in the Textiles Industry

Basel Younes

Abstract

The current chapter is focused on biopolymers and Bionanocomposite as environmentally friendly materials, modeling of the production processes, and coating of bio-textiles. Different industries use biopolymers and Bionanocomposite in for the current environmental applications. Furthermore, composition and classification of biopolymers, the theoretical methods, and factorial experimental designs (FED) for optimization and modeling processes of the environmentally friendly textiles used as an alternative to traditional chemical textile products with zero to low environmental footprint are studied at acceptable cost. This chapter will also describe the novel optimization, experimental factorial design, and how the novel modeling methods will help less experienced polymer designers in taking the best experimental decision controlled by the design factors. It also discusses how the fully biodegradable polymers support the industry by decreasing the processing energy, material and manufacturing costs. Finally there are an overview of the current and future developments of biodegradable polymers applications in modern bio-textiles industries.

Keywords: biodegradable textiles, biopolymers, coating, textiles industry

1. Introduction

The worldwide production of non-renewable textiles has increased, and textile producer are searching for solutions to overcome the waste problem [1]. There is a new generation of bio- textile materials based on petroleum, animal sources or agricultural [2] with a reasonable solution [3]. Micro-organisms (such as fungi and bacteria) broke down the biopolymers in the environment; such as a cotton fiber [4].

Table 1 [5] illustrates how degradation results are dependent on biodegradation type, the converted substrate, and residue makeup, with full degradation. Biodegradation happens within the biosphere, the organic chemicals are changed to simpler compounds and mineralized.

Researchers have established many standardized testing procedures for the evaluation of the compost-ability and biodegradability of polymers using mixed cultures [6–11]. Others have studied the effects of blend ratios on the degradation process on biopolymers [12–14]. The life cycle analysis is one of the methods simulating the development of biopolymers, with green fibers having a shorter life cycle than those that are oil-based. A green life cycle is given in **Figure 1** [15].

IntechOpen

Original substrate	Biodegradation Type	Converted into
Polymer	Aerobic	CO2& H2O & (Biomass & Residue)
	Anaerobic	CH4& CO2& H2O & (Biomass & Residue)

Table 1.
Degradation results depending on biodegradation type [5].

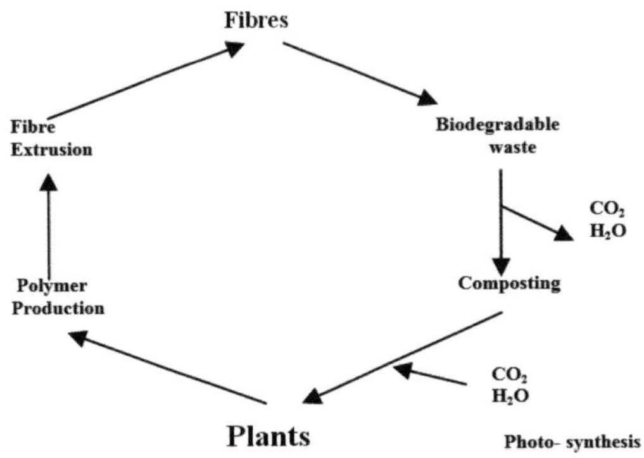

Figure 1.
Life cycle of compos-table biodegradable fibers [15].

Biodegradable polymers are smart polymers which are currently being used in many fields such as tissue culturing, biomedical, agriculture, food and intelligent textiles [16, 17]. Considering environmental hazards [18], the main factors controlling the market scope and size of biodegradable polymers are material properties and cost [19]. Plasticizers are added to biopolymers during the extrusion process to decrease the intermolecular hydrogen bonds, to limit microbial growth, and to stabilize product properties [20]. The degradation rate of blended bio-polymers set the degradability of the produced mix [21]. Bio-materials take a key role in the of nanotechnology improvement as friendly materials. Nanomaterials have attracted considerable attention in medical delivery applications [22]. Classification and composition of biodegradable polymers will be briefly discussed; their applications in modern bio-textiles as well as textiles. Furthermore, modeling of biopolymers' melt spinning process and factorial experimental design, optimization of the production processes for intelligent bio-fibers via statistical experimental design (SED), forecasting program for the fiber extrusion, as well as the future applications of biodegradable polymers in the modern textiles industry are also presented.

Electro-spinning of biopolymers has gained substantial attention in the last two decades, triggered mainly by the potential applications of electro-spun nanofibers in nanoscience and nanotechnology for tissue engineering [23]. Tissue engineering is a advanced technology, electrically conductive biodegradable composites are used in tissue engineering and bioelectronics [24].

2. Biodegradable polymers, classification and composition

Biopolymers are made from the agro-polymers (starch and cellulose), or are obtained by microbial production such as the polyhydroxyalkanoates. In polyhydroxybutyrate production, sugarcane, mustard, switch grass and corn have been recognized as candidates for genetic modification. Some polyhydroxybutyrate types polymerized chemically from agro-resources or chemical synthesis [25]. Classification of biopolymers and their origins are listed in **Table 2** [3].

Biodegradable polymers are produced from aliphatic (linear) highly amorphous, flexible polymers and aromatic rings semi-crystalline, rigid polymers. The classification, development and synthesis of the main bio-based polymer types from biomass and microbial production, or from renewable resources are listed in **Figure 2** [26, 27]. Aliphatic-aromatic copolymers can be synthesized and used in biomedical and agricultural applications by employing non-woven technology to produce products such as disposable wipes, refuse bags, seed mats and erosion control items [28, 29].

Biopolymers classification			
Natural		**Synthetic**	
Proteins	Gelatine, casein, silk and wool	Aromatic Polyesters	Polybutylene succinate terephthalate
Polysaccharides	Starch, cellulose, lignin and chitin	Aliphatic Polyesters	Polyglycolic acid, Polybutylene succinate and polycaprolactone
Lipids	Castor Oil,Plate Oil and animal fats	Aliphatic-Aromatic co-Polyesters	
Polyesters 1- micro-organism or plants	1-polyhydroxyalcanoates, poly-3-hydroxybutyrate	Polyvinyl-alcohols	
2- bio derived monomers	2- polylactic acid	Modified Polyolefin	Polyethylene or polypropylene & specific agents

Table 2.
Classification of biopolymers [3].

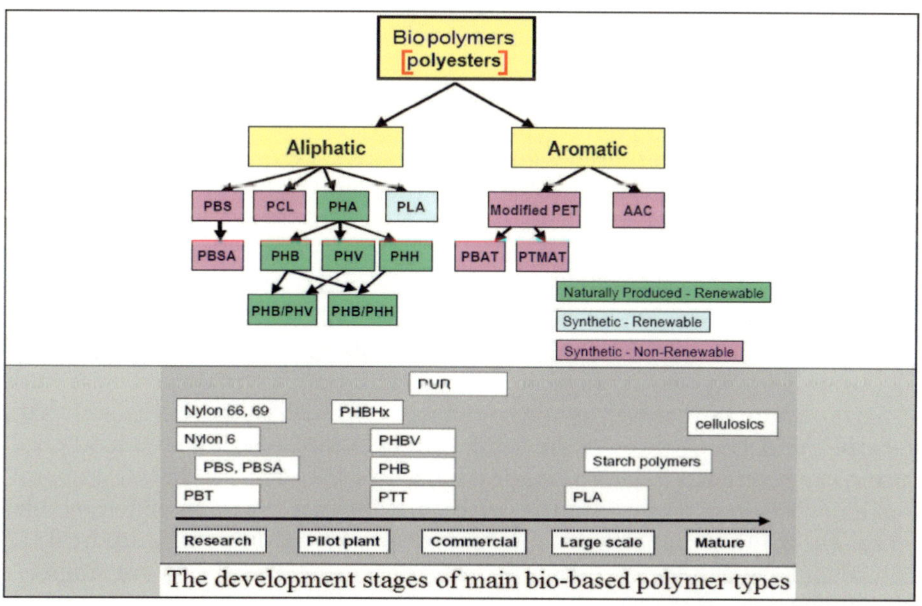

Figure 2.
Classification and development of bio-based polymers [26, 27]. ***PCL – polycaprolactone; PBS – polybutylene succinate; PTT– polytrimethyleneterephthalate; PBSA – polybutylene succinate adipate; PHH – polyhydroxyhexanoate; AAC – Aliphatic-Aromatic Co-polyesters; PHV – polyhydroxyvalerate; PET – polyethylene terephthalate; PLA – polylactic acid; PBAT – polybutylene adipate/ terephthalate; PHA – polyhydroxyalkanoates; PTMAT– polyethylene adipate/terephthalate; PHB – polyhydroxybutyrate (type of PHA); PBT– polybutylene succinate; PHBV, PHBHx- types of PHA; PUR– polyurethanes.***

3. Techniques for the preparation and synthesis of biopolymers

Due to biopolymers' biodegradability and substantially, their applications are environmentally friendly [30]. Biopolymers properties influence the shelf-life and the product's biodegradability. The possibility of increasing the strength of bio-polymers and bio-composites have been studied without decreasing the biodegradability [31].

Poly (butylene succinate-co-butylene terephthalate) co-polyesters have much better thermal stabilities in nitrogen compare to air [32]. It was reported that "Poly(butylene succinateco- ethylene succinate-co-ethylene terephthalate)"can be polymerized from three pre-polymers of ethylene succinate, butylene succinate, and ethylene terephthalate by direct poly-condensation [33]. An ideal random copolymer (Poly(butylene terephthalate-succinate-adipate) from aliphatic units (BA and BS) has a rubber-like tenacity curve [34]. Aliphatic aromatic copolyester (AAC) could potentially modify the basic BTA (1,4-butanediol, adipic acid, and terephthalic acid) structure and may become commercialized [35, 36]. Development of biodegradable aliphatic-aromatic co-polyesters began with the study of different modes of degradation [37, 38]. Aliphatic biopolymers are biodegradable and sensitive to hydrolysis; their flexible chain fits easily into the active site of an enzyme [39]. Aromatic biopolymers have favorable physical properties such as resistance to bacterial, fungal and hydrolysis attack [40] but degrade if they are co-polymerized with aliphatic bio-polymers [41]; breaking down by means of hydrolytic or/and enzyme degradation [42]. It was reported that inclusion and/or incorporation of aromatic monomer groups in the aliphatic polyesters' main chain, can potentially enhance their mechanical properties [43]. The randomness and the length of the polymer chains aid in understanding the biodegradation behavior for aliphatic-aromatic co-polyesters [44]. Polyester-based nano-particulates could be easily prepared by solvents diffusion or evaporation methods. The degradability of the oligomers would decrease by increasing chain length [45], thus the amorphous part of the polymer would become that which is degraded [46].

4. Biopolymer and industrial applications

Biodegradable polymers used widely in industries such as textiles, packaging, fast-food container and packaging, paper coating, agriculture mulch films, medical products, tubes, lawn and garden waste bags, disposable wipes, erosion control, biologically-based resins, car parts, glass fibers agents, as well as coatings and adhesives [3, 47, 48]. Various blending ratios of regular and waxy corn starches with co-polyester were extruded into loose-fill foams [49]. PCL (polycaprolactone) are used to made spun fibers, scaffold fibroblasts and myoblasts for soft tissue engineering [15]. Natural biopolymer-based films and the packaging materials have been studied [50, 51]. Ochratoxin-A as well as amycotoxinis a common food contaminant that enters the human body through the consumption of improperly stored food products and can be used as a electrochemical biosensor [52]. Polysaccharide and protein based biopolymers can be utilized as coatings to enhance the fruits and vegetables quality; The medical biopolymers applications are [53] include extracorporeal (i.e., artificial kidneys, fluid lines, dialysis membranes, catheters, wound dressings, artificial skin, etc), temporary implants (i.e., degradable sutures, as well as arterial stents, tissue/cell transplants' scaffolds, temporary vascular grafts, etc), and permanently implanted devices. Biopolymer fibers with typical morphology find applications in bone tissue engineering [54] and as a degradable nano-fiber [55]. Wound dressing materials must be biocompatible, anti-bacterial, prevent infection, and provide a suitable moist environment [56, 57]. Chitosan complexed with gelatin has been useful as a surgical dressing at a ratio of 3:1 (chitosan:gelatin), as it stimulates hemostasis and accelerates tissue regeneration [58, 59]. Their fabrication provides appropriate biodegradability and excellent cell adhesion activity, both useful in making a novel and elastomeric bioactive vascular tissue scaffold [60]. A fiber based on chitosan and starch which was loaded with drug has had successful applications in drug delivery [61]. Various fabrication methods have been employed in the preparation of bio-polymeric membranes and a film-casting process [62, 63].

5. Biopolymers and biodegradable textiles industry

The textile industry has played a important role in the exchanging of goods and impacted by novel techniques through the development of more environmentally friendly processes [64–67]. Many biodegradable fibers may be natural, regenerated or synthetic such as Ingeo (Natureworks), LLC produced from corn, a biodegradable thermoplastic polyatide (PLA); Lenpur produced from wood pulp of harvested white pine tree clippings; and Modal and Tencel/Lyocell produced by Lenzing from wood pulp of beech and eucalyptus trees, and biodegradable aliphatic/aromatic multi-block co-polyesters. The largest application of alginates in textiles can be found in textile printing, the spinning and weaving of temporary fibers from calcium alginate [15]. Bio-based fibers like X-Static, Meryl Skinlife, Diolen Care, Trevira Bioactive are enriched with innovative antimicrobial products such as technical products, working uniforms, sportswear [68].

5.1 Fibers industry and biodegradable polymers

The melt spinning processing technologies with availability of biodegradable materials along have aided in qualitative and quantitative improvements [69]. The fibers reinforcing improve the relationship between the process parameters and the material properties [70]. Many natural fibers added to biopolymers as reinforcements (flax, cellulose acetate, bamboo, pineapple, ramie, kenaf, henequen, jute, sisal, and hemp); improving the strength without affecting the biodegradability [71]. Fibers of the poly(β-hydroxybutyrate) were produced via multistage melt-extrusion as well as gel-spinning [72]. The dry-jet wet spinning method was used to extrude the cellulose/NH3/NH4SCN solution [15]. By "dry-jet wet spinning" and using a cellulose/hydrolyzed starch-grafted-polyacrylonitrile solution, the mechanical properties of Lyocell fibers were improved [73]. An extruded Lyocell fibers were reported have potential uses in filters, geo-textiles, surgical gauze [74]. Cellulose fibers are used for the design of intelligent, bioactive, and biocompatible composites [75]. Preparing sol–gel derived biodegradable SiO2 gel fibers [76] for drug release consists of three steps: an initial burst, followed by a diffusion controlled release behavior, and finally a step with a slower release rate. By incorporating responsive hydrogels in textiles, the surface energy switches between hydrophilic/hydrophobic, with the results listed in **Table 3** [77].

There is good compatibility between the chitosan obtained from shrimp shells and starch-based polymers when forming a chitosan/starch fiber. Some researchers have made fiber from starch and biodegradable glycerine-based polymers with/without PLA and glucitol, while others extrude high strength PLLA fibers. The type LA 0200 K of PLA is processed at a high speed spinning (draw ratio = 6)in a spin drawing [78]. A fiber from soya bean protein is crafted by forcing a globular protein to become a fiber forming protein; the fiber has to be cross-linked if fibrous products are to be obtained [79]. Fibrous materials are segregated into two basic groups, the first can be placed on the surface of materials (i.e., surgical covers, gauzes, diapers and tampons), and the second can be placed inside an organic tissue (i.e., surgical threads, tendons and ligament implants, meshes, stents, and vascular grafts).

Biopolymer	Monomer	Natural origin	Production Micro-organism	Hydrogel	Fiber	Commercialized
Cellulose	glucose	Plants, Micro-organism	A. xylinum	—	Yes	Wound covering
Hyaluronic acid	N-acetyl-glucosamin	Vertebrats, Streptococci	Streptococci, *B. subtilis*	Yes	No	Dermal fillers, Visco supplementation
γ-PGA	Glutamic acid	Bacillus spp., Micro-organism	Bacillus spp.	—	No	Cosmetics, fertilizer, flocculant
Silk	protein	Bomby-xmorii, spider, bee	*E. coli*,	Yes	Yes	Cosmetics, cell adhesion scaffolds
Collagen	protein	Vertebrates	Yeasts	Yes	Yes	Scaffolds, Cosmetics
Chitosan	N-aeetyl-glucosamin (partly deacylated)	fungi	Yeast, bacteria	Yes	Yes	Flocculant, Filter membranes, Cosmetics, wound covering

Table 3.
Monomers, origin and fiber formation of smart biopolymers [77].

5.2 Fabric industry and biodegradable polymers

Weaving, knitting, nonwoven web forming (carding, spun-bond and wet-laid) and nonwoven-bonding (stitch-bonding, needling, calendaring and hot air bonding) are fabric forming technologies. Biodegradable non-woven webs and disposable articles contain fibers such as cotton, hemp, milkweed floss, flax fiber, wool, silk, chitin and chicken feathers [80]. There are many examples in biodegradable nonwoven such as biodegradable cotton-based nonwovens (cotton/cellulose, or cotton/ biodegradable co-polyester) and (PTAT co-polyester and PLA). Cotton/(co-polyester/PP) nonwovens along with absorbency and flexural rigidity have suitable mechanical properties and they are better than that of cotton/co-polyester nonwovens [81–83].

Sanitary and medical textiles, geotextiles, filtration media, within the automotive industry, PLA based hair caps, Bionolle 3001 nonwovens, Landlok biodegradable erosion control coconut fibers mats, Kenaf fiber nonwovens, refuse bags, drain filters made from fine denier PLA nonwovens, and biodegradable filter materials are used for both air and liquids. A biodegradable thermoplastic polymer and a plasticizer could be used to produce a starch matrix of the finely attenuated fibers which could have applications as environmentally degradable nonwoven webs and articles [84].

Researchers have produced biodegradable cotton-based, nonwovens by using blends of cotton, flax and biodegradable thermoplastic fibers that act as binders [85]; Biodegradability was monitored, with 40% of the initial weight lost after 8 weeks composting [86]. To reduce the cost, researchers have made a pure nonwoven material from co-polyester by a direct melt-blowing process [87]. Woven tubes (3 to 6.5) mm are developed using Polyglactin 910 biodegradable yarn on a narrow width loom [88].

Biodegradable poly (L-lactide-co-caprolactone) fabrics of nano/micro- structured can be made Using CH2Cl2 as a solvent in electro-spinning. The electro-spun elastomeric nano-fiber fabric is used as a functional scaffold in tissue engineering (i.e., cardiovascular, muscular) [89]. The Belgian Textile Research Centre's projects include: Noterefiga for bio-based comfort textiles, Bioagrotex for agro-textiles (agriculture, horticulture, gardening and construction), Green-Nano-Mesh for medical areas, Dura cover for woven PLA taped ground covers, Hortaflex and Weed Control for PLA based nonwovens, and the BiobasedFilbio project for knitted PLA insect screens for climate control.

There are various commercialized fabrics made from naturally derived biopolymers such as those found in the Ethical Fashion Forum in London: POLY Acid Ingeo bio fibers; QMilchfibers, Lenzing's Modal fiber, Micro Modal fiber, Lyocell fiber, POLARTEC polyester, unique corn-based PLA fleece; and Cork shell made from cork to form high quality textiles for lightweight spring and summer jackets. Biopolymers based in intelligent and/or stimuli responsive polymeric systems have been developed and reported by researchers for the functional finishing of textiles [90]. Scientists proposed changes of the polymer backbone in a reversible formation of PLA-dye complexes [91]. Sorona is used in the coat fabric for jackets, trench coats and outerwear with its 37% renewably sourced plant-based components; they lose their wrinkles with one quick snapping motion. Bio-based fabrics made of wool and "BIOPHYL" or "TENCEL". Some commercial products are made from spider silk [92] and could be used in the bullet-proof vests industry [15].

5.3 Fiber and fabric coatings and biodegradable polymers

Modification techniques of the biopolymer's surface includes coating, oxidation by low-temperature plasma, and surfactant addition blending with various derivatives [93]. Cyclodextrins or linear carbohydrate biopolymers were attached to the textile to allow frequent use and washing [94]. Regenerated cellulose fibers were treated by plasma activation using a chitosan solution [95]. Cellulose was coated by chitosan nano-particles to reduce the cost and non-toxic methodology [75]. While studying their development as well as characterization, both the organic cotton based bandages and cotton were coated separately on the gauze structure using chitosan-sodium alginate polymer, calcium-sodium alginate polymer and subsequent mixtures of the two, thereby improving its antibacterial and wound healing properties [96]. For dyeing and printing, the dextrin derivative surfactant improves the whiteness and wetting properties of cotton fabrics [97]. The chemical surface treatments of jute fabrics involve bleaching, dewaxing, cyanoethylation, alkali treatment and vinyl grafting are used as reinforcing components in biodegradable matrix composites, which are environmentally friendly materials [98]. A Knitted Dacron graft made of polyethylene oxidepolylactic acid were coated with a polymeric biodegradable sealant [99]. Layer-by-layer electrostatic deposition is used to coat the material by adding dextran sulphate and chitosan to a soybean based polymer [100]. The functional finishing of the micro- and nano-sized hydrogels improve response times [101].

Nano-composites and nano-structured coatings improve mechanical strength and flexibility, temperature and moisture stability, as well as durability. "Metal Rubber"(Nano Sonic Inc.) combines the rubber and metal properties, and it is used in artificial muscles, electrically charged aircraft wings, and protective biopolymer clothing [102]. Hydrogel-based biopolymers are used for the functional finishing of textiles by surface modifying systems [103].

6. Case study: modeling of biopolymers' melt spinning process

All the production process parameters must be controlled to ensure the quality and then the significant main factors must be analyzed [104]. Commercially, it is a challenge to develop a new competitive product [105]. Some research is based on statistical analysis, mathematical simulation and modeling of the processes of fiber formation, and examples of their post-processes have been reported in literature [106–116]. The practical software-based approach has improved the confidence benefits of experimental design and simulation [117]. **Figure 3** shows a flow chart for the methodologies used for obtaining the program, starting from the data and statistical modeling methods and SED. Online quality control tools were utilized for prediction, measurement, correction as well as adjustment and feedback [118].

The aliphatic aromatic co-polyester fibers extrusion process was investigated in this work, and statistically modeled [119]. A linear biodegradable oil-based polymer (LAAC-flexibility component of Solanyl) and branched aliphatic-aromatic co-polyester (BAAC-Ecoflex F BX 7011) were used to study the effects of the extrusion process and the properties of fibers. The study describes the melt spinning of aromatic-aliphatic co-polyester depending on the extrusion thermal profile effect on as-spun fiber properties. The molten material flowed easily when the viscosity decreased and smoother extrudates were obtained at shear rates greater than 4.5 s-1 [120].

6.1 Factorial experimental design for melt spinning of biodegradable fibers

Factorial experimental design provides data about the optimization of the average response values in regards to the factor levels [121]. The STATGRAPHICS program is used to design the experiment random order matrix and to simulate the main data in one block experiments.

The studied factors for the fiber extrusion process include: speed of spin finish, quenching air speed, metering pump speed, and winding speed, as well as, melt-spinning or extrusion temperature. The analyzed levels of each parameter were listed in **Table 4**; the thirty-two trials matrix for the five control factors was applied for as-spun fibers analysis. **Figure 4** shows an SEM photomicrograph of the cross-section and surface of the fibers; fibers had an acceptable uniform surface and possessed a uniform circular cross section.

In this case study [122–124], several statistical tools were utilized for statistical analysis including the surface plot, normal probability plot, the main effect plot, pareto chart, interaction plot, as well as analysis of variance (ANOVA). Implementation of forecasting statistical methods plays a major role in creating a planning program and

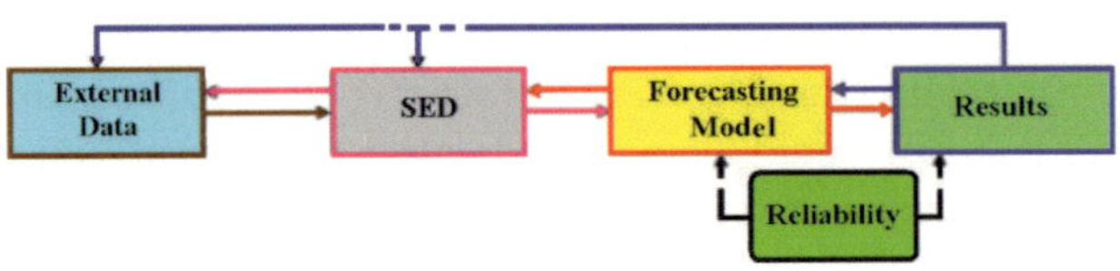

Figure 3.
The flow chart of the statistical method.

Factor abbreviation	Factor name	Level	
		Low	**High**
T	Melt-Spinning Temperature, °C (LAAC)	**130**	**145**
	Melt-Spinning Temperature, °C (BAAC)	**145**	**160**
MPS	Speed of Metering Pump, rpm (2.4 cc/rev))	**6**	**12**
QA	Speed of Quench Air, % (velocity m/sec)	**35**	**50**
SF	Speed of Spin Finish Pump, rpm (0.15 cc/rev)	**0.35**	**0.50**
WS	Winding Speed, m/min	**50**	**100**

Table 4.
Factors and the selected levels for the spinning experiments of as-spun fibers.

a plan for the production process regression. A detailed experimental arrangement of the calculated results of spin draw ratio, birefringence, drawability, die head pressure, crystallographic order as full-width half-maximum (FWHM), filament temperature averages, count, tensile properties, diameter, and thermal shrinkage was completed. According to the drawability characterization, biodegradable fibers (i.e., as-spun) should consist of a drawn construction and be conducive to orient along the fiber axis of the chain [125]. There is a clear relationship between the draw down ratio and the orientation of the fibers and having a significant effect on the drawability. In other words, the overall orientation of fibers was increased and the draw ratio decreased as the spin draw ratio increased. Temperature significantly influenced the spin (down) draw ratio and fiber drawability that affects the flow rate and tension value. To study the effects of the factors as well as their statistical significance an ANOVA study was conducted. A factor was considered to have a significant effect if the F ratio (an ANOVA statistic) was shown to be more than the statistical value (F /4.49/at the appropriate level α =0.05) or had P-value smaller than 0.05. The ANOVA results from the experiments are presented in **Table 5**.

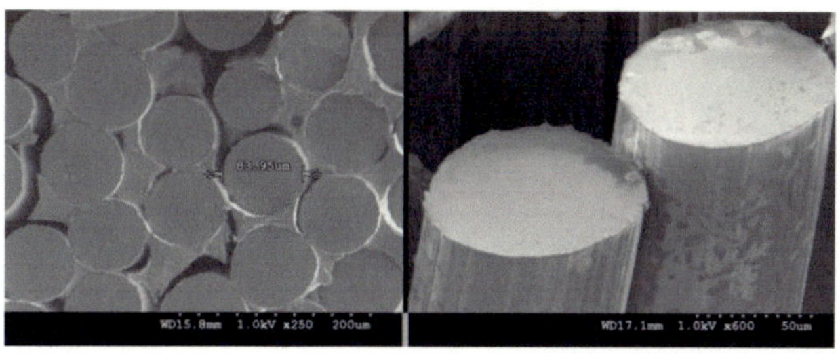

Figure 4.
The surface and cross section of the biodegradable fibers [119].

Source	Drawability	
	F	P
T	19.7	0.000
MPS	112.1	0.000
QA	0.2	0.666
SF	0.3	0.599
WS	120.5	0.000
T &MPS	0.1	0.737
T & QA	0.2	0.666
T& SF	0.1	0.810
T& WS	7.2	0.016
MPS& QA	2.0	0.176
MPS& SF	0.2	0.666
MPS& WS	1.5	0.240
QA & SF	0.0	0.885
QA & WS	0.0	0.962
SF& WS	0.3	0.599

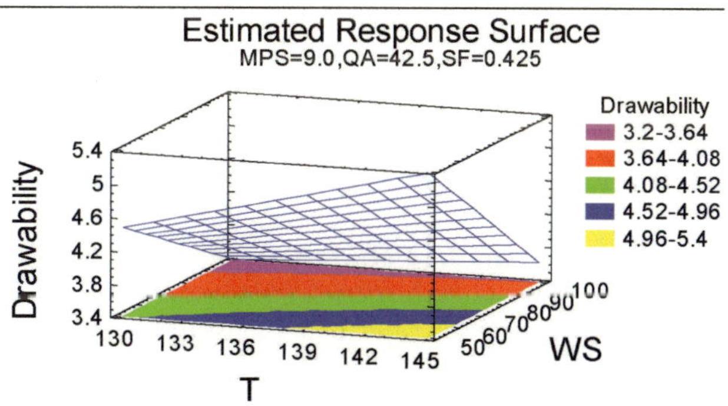

Estimated response surface for the drawability
(T& WS)

Estimated response surface for the drawability
(T& WS)

Table 5.
ANOVA results of factor effects on the drawability.

The significance of factors were PWS > PMPS> PT in the drawability analysis, while no significant effect was observed due to other factors. The P-value (0.016) of T&WS is lower than 0.05 and therefore is significant. The most significant factors were T, MPS and WS. Metering pump speed was observed to have interaction with winding speed; the speeds' relationship oriented the fiber chains as well as added different spin draw ratio, having an effect on drawability later. Multiple and individual regressions optimized for the quality required for various applications and identified the factors' effects and interactions to determine the direction of those that are significant by using the estimated response surfaces. A twist was observed in the 3D surface response diagrams for T and WS (**Table 5**), thus the interaction is significant and agrees with the previous statistical results. This interaction will affect the structure of the as-spun fibers and help to extend the chains to achieve high orientation along the axis of fiber. The regression Eq. (1) was obtained from the analysis and forms the simplified models of the experimental data (coded values in **Table 4**). The regression equations forecast the fiber properties and accurately predict the properties in the final fibers produced. The mathematical regression model forms one of the basic source codes in the designed forecasting application, which will present the extrusion of aromatic-aliphatic co-polyester fiber.

$$\begin{aligned} \text{Drawability} = {} & a + b*T + c*MPS - d*QA - e*SF + f*WS \\ & + b1*T*MPS + b2*T*QA + b3*T*SF - b4*T*WS \\ & - b5*MPS*QA - b6*MPS*SF - b7*MPS*WS \\ & - b8*QA*SF + b9*QA*WS - b10*SF*WS \end{aligned} \tag{1}$$

Where: a, b, c, d, e, f b1–10, are statistical constants for the drawability calculated by the **STATGRAPHICS** program.

They were also affected by high extrusion speed at which the shear rate affects the morphological structure [126]. Employing the same technique, the overall orientation, spin draw ratio, crystallographic order, die head pressure, diameter, tensile properties, thermo-graphic measurement and thermal shrinkage were also analyzed and modeled. The statistical analysis models simulated the significant factors, their interactions, and gave useful results with some expected outliers which could be due to experimental and/or testing errors.

6.2 Forecasting program for the fiber extrusion

In the programming process, the relationship between the key inputs (factors) and the performance measures (responses) using factorial statistical experimental design technology are reported. The statistical data and regression formulas are represented as a computer application. Microsoft Visual Basic was used to write a forecasting program that could be utilized for the as-spun AAC fibers' extrusion process. The program offers the management of regression models for responses based on statistical factorial design, design analysis and process simulation. Conversion and summarization of the C++ source code into a simple flow chart was completed (**Figure 5**).

After selecting the polymer grade, the program requests the parameters' values, calculates the values' responses by using regression equations and then gives the results. The data from the input conditions was used to obtain the structural, mechanical and physical data. The program was designed as two windows. The first window is the input window for process conditions (**Figure 6**); the second interface is the output result window (**Figure 7**).

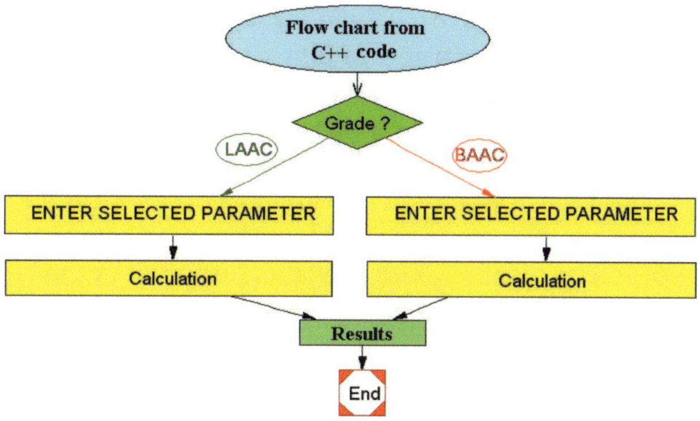

Figure 5.
Schematic program process.

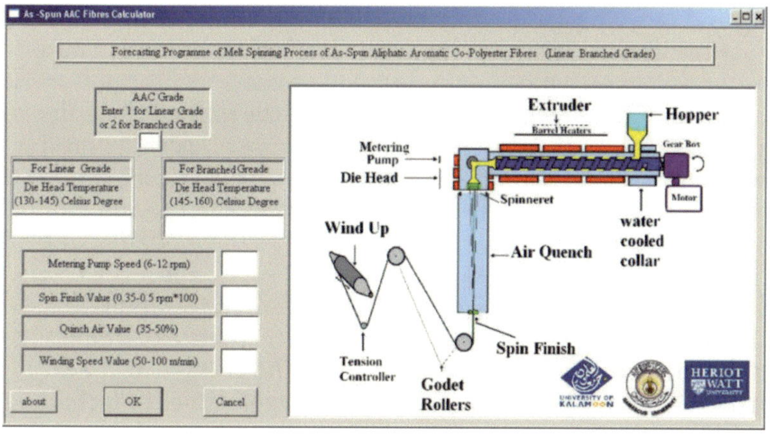

Figure 6.
The main input interface/window for process conditions input.

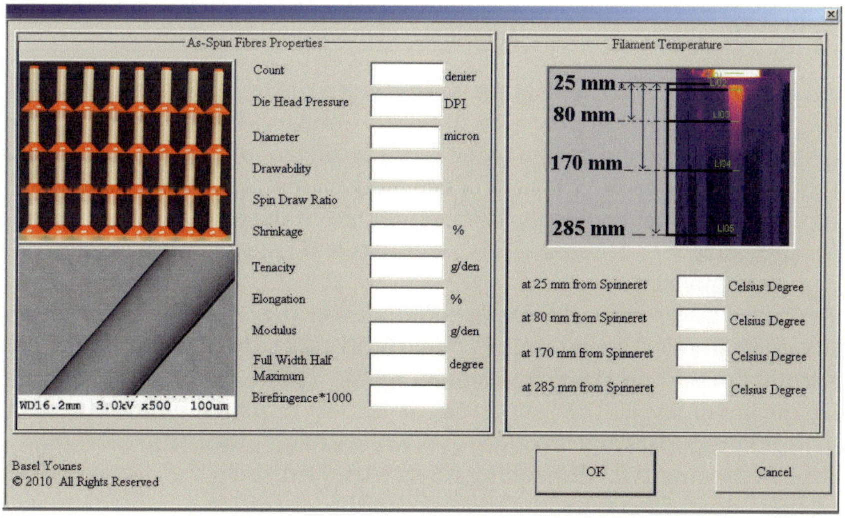

Figure 7.
The output interface/window for filament temperature in the machine's cooling window and the fiber's structural, mechanical and physical properties.

Each factor is represented as a record and it may be owned by more than one record, leading to a network-like structure. The multiple regression analysis and previous forecasting models provide a basis for identifying the relationship between process-input and process-output data; and formation of a source code to be used in the forecasting program. The 30 hole spinneret (diameter is 0.4 mm, l/d ratio is 1.2) was used. The programmed application powerfully supports product development, design process control, quality assurance and product performance evaluation; it displays data on the screen or sends data to a file or other devices. **Figure 7** shows the output interface/window for filament temperature in the machine's cooling window and the fiber's structural, mechanical and physical properties. Each factor is represented as a record and relationships between other factors through a matrix design.

7. Textiles industry future and biodegradable polymers

Reusing and recycling properties are the main goals for the environmentally friendly textile industry [127]; saving time, cost, and materials. There are various methods in which bio-polymers are processed, such as polymerization, crystallization and manufacturing depending on the polymer's nature and application [128]. Biomechanical engineering design of clothing products is still at the development stage. Biological health and psychological happiness are critical indexes reflecting quality of life. Bio-material processing and bio-garment simulation has grown from basic shape modeling to the modeling of cloth with complex physics and behaviors. Bio-textile engineering approach offers precise details of modeling cloth at a micro, the graphics software simulate the fabric to form the final deformation and animation. The acceptance of biopolymer materials as commonplace in textile industry require the passage of time. Biopolymers boast greater environmental consciousness than most modern technologies, as they have the potential to significantly reducing cost, energy and materials for future generations. Modeling crosses the boundaries of academia, science and industry [129].

8. Conclusion

This chapter reviewed the relationship between the bio-polymers and the modeling of the production processes in textile industry. The theoretical techniques and factorial experimental design together with the biopolymers' classification and preparation methods open new field of the modern textiles, from fiber to fabric forming and coating technologies. Biopolymer use in the textile industry is an exciting and innovative area of research for scientists and researchers alongside textile and polymer engineers.

In the case study, results obtained should answer the fairly complex demands posed by multi-applications running concurrently with the application programs (or processes) in the computer. It is limited by the regions of the studied factors between the factor levels. The program's results help in achieving a balance between the enhanced properties and the fiber cost while saving processing cost, material and the power required for enhancing fibers. After finishing the processes for modeled biodegradable fibers, the process conditions (process-input data) selected depends upon the user. The produced environmentally friendly, economical, energy saving fibers can be potentially utilized in textiles, agricultural, as well as horticultural applications.

Acknowledgements

I wish to express his deep appreciation to Mechatronics Department, Engineering Faculty, Kalamoon University &Textiles Department, FMEE, Damascus University.

Author details

Basel Younes
Faculty of Engineering, Kalamoon University, Damascus, Syria

*Address all correspondence to: basel.younes@uok.edu.sy

IntechOpen

© 2021 The Author(s). Licensee IntechOpen. This chapter is distributed under the terms of the Creative Commons Attribution License (http://creativecommons.org/licenses/by/3.0), which permits unrestricted use, distribution, and reproduction in any medium, provided the original work is properly cited.

References

[1] Chandra R, Rustgi R. Biodegradable Biopolymers. Great Britain: Elsevier Science Ltd; 1998.

[2] Steinbüchel A. Biopolymers Vol. 10: General Aspects and Special Applications. Germany: Wiley-VCH; 2003.

[3] Smith R. Biodegradable Polymers for Industrial Applications. England: Woodhead Publishing Limited; 2005.

[4] Harzallah O, Drean J-Y. Macro and Micro Characterization of Biopolymers: Case of Cotton Fibre. In: Elnashar M, editor. Biopolymers. Croatia: Sciyo; 2010. p. 193-218.

[5] Bastioli C. Handbook of Biodegradable Polymers. Shawbury, UK: Rabra Technology; 2005.

[6] Standard guide for assessing the compostability of environmentally degradable plastics, D 6002-96. American Society for Testing and Materials, Washington, DC. 1996.

[7] Sawada H. ISO standard activities in standardization of biodegradability of plastics-development of test methods and definitions. Polymer degradation and Stability. 1998;59:365-70.

[8] Pavlov MP, Mano JF, Neves NM, Reis RL. Fibres and 3D mesh scaffolds from Biodegradable starch based blends: production and characterization. Macromolecular bioscience 2004;4 776-84

[9] Chen Y, Wombacher R, Wendorff JH, Visjager J, Smith P, Greiner A. Design, Synthesis, and Properties of New Biodegradable Aromatic/Aliphatic Liquid Crystalline Copolyesters. Biomacromolecules. 2003;4(4):974-80.

[10] Han L, Zhu G, Zhang W, Chen W. Composition, Thermal Properties, and Biodegradability of a New Biodegradable Aliphatic/Aromatic Copolyester. Journal of applied polymer science. 2009;113(2):1298-306.

[11] Averous L. Biodegradable multiphase systems based on plasticized starch: a review. Journal of Macromolecular Science - Polymer Reviews. 2004;44(3):231-74.

[12] ASTM. D 5338, Test Method for Determining Aerobic Biodegradation of Plastic Material under Controlled Compositing Conditions. 1998.

[13] Sami J. Processing of the 4th Conference on Biogically Degradable Mterials. Presentation No D 95 1999 (Festung Marienberg, Würzburg, Germany).

[14] Biresaw G, Carriere CJ. Compatibility and mechanical properties of blends of polystyrene with biodegradable polyesters. Composites: Part A. 2004; 35:313-20.

[15] Blackburn RS. Biodegradable and Sustainable Fibres. Cambridge, UK: Woodhead Publishing; 2005.

[16] Holding W. Biodegradable Polymer Supply Chains: Implications and Opportunities for Australian Agriculture. Australia Rural Industries Research and Development Corporation; 2004.

[17] Monsanto. Biopolymers to Give Cotton Fibers Synthetic-Like Qualities. U.S.A: Monsanto company; 2003.

[18] Thakore SI. Role of Biopolymers in Green Nanotechnology. In: Verbeek J, editor. Products and Applications of Biopolymers. Croatia: Tech; 2012. p. 119-40.

[19] Kalambur S, Rizvi SSH. Biodegradable and Functionally Superior Starch-Polyester Nanocomposites from Reactive Extrusion. Journal of Applied Polymer Science. 2005;96(4):1072-82.

[20] Crank M, Patel M, Marscheider-Weidemann F, Schleich J, Hüsing B, Angerer G. Techno-economic Feasibility of Large-scale Production of Bio-based Polymers in Europe (PRO-BIP)-Final Report Utrecht University& Fraunhofer ISI,Spain 2004.

[21] Amass W, Amass A, Tighe B. A review of biodegradable polymers. Polymer international. 1998;47:89-144.

[22] Singha K, Namgung R, Kim WJ. Polymers in Small-Interfering RNA Delivery. Nucleic Acid Therapeutics. 2011;21(3):133-47.

[23] Wanga X, Ding B, Li B. Biomimetic electrospun nanofibrous structures for tissue engineering. Mater Today (Kidlington). 2013;16(6):229-41.

[24] Shi G, Rouabhia M, Wang Z, Dao LH, Zhang Z. A novel electrically conductive and biodegradable composite made of polypyrrole nanoparticles and polylactide. Biomaterials. 2004;25:2477-88.

[25] Takiyama E, Fujimaki T. Biodegradable Plastics and Polymers. Amsterdam: Elsevier; 1994.

[26] Nolan-ITU. Environment Australia:Biodegradable Plastics-Development and Environmental Impacts. Nolan-ITU, East Kew, Victoria. 2002.

[27] Crank M, Patel M, Marschcider-Weidemann F, Schleich J, Hüsing B, Angerer G, et al. Techno-economic Feasibility of Large-scale Production of Bio-based Polymers in Europe: Technical Report EUR 22103 EN. European Science and Technology Observatory. 2005;European Communities.

[28] Ganesh K, Alp AH, Russell GR, inventorsBiodegradable Aliphatic-Aromatic Copolyesters, Methods of Manufacture, and Articles Thereof 03/24/2011.

[29] Eastman polymers for fibres. Eastman chemical company, USA2002.

[30] NIIR-board. The Complete Book on Biodegradable Plastics and Polymers (Recent Developments, Properties, Analysis, Materials & Processes). Delhi, India: Asia Pacific Business Press Inc.; 2006.

[31] Stevens ES. Green Plastics, Plastics and the Environment USA: Princeton University Press; 2001.

[32] Li F, Xu X, Li Q, Li Y, Zhang H, Yu J, et al. Thermal degradation and their kinetics of biodegradable poly(butylene succinate-co-butylene terephthate) s under nitrogen and air atmospheres. Polymer Degradation and Stability. 2006;91:1685-93.

[33] Deng L-M, Wang Y-Z, Yang K-K, Wang X-L, Zhou Q, Ding S-D. A new biodegradable copolyester poly(butylene succinate-co-ethylene succinate-co-ethylene terephthalate) Acta Materialia. 2004;52(20):5871-8.

[34] Shi XQ, Aimi K, Ito H, Ando S, Kikutani T. Characterization on mixed-crystal structure of poly(butylene terephthalate/succinate/adipate) biodegradable copolymer fibers. Polymer. 2005;46:751-60.

[35] Rolf-Joachim M, Ilona K, Wolf-Dieter D. Biodegradation of polyesters containing aromatic constituents. Journal of biotechnology. 2001;86(2):87-95.

[36] Witt U, Muller R-J, Deckwer W-D. New biodegradable polyester-copolyesters from commodity chemicals with favorable use properties. Journal of Polymers and the Environment. 1995;3(4):215-23.

[37] Park SS, Chae SH, Im SS. Transesterification and crystallization behavior of poly (butylene succinate)/ poly(butylene terephthalate) block

copolymers. Journal of Polymer Science, Part A: Polymer Chemistry 1998;36(1):147-56

[38] Hayes RA. Aliphatic-aromatic copolyesters. US Patent 6485819. 2002.

[39] Pan P, Inoue Y. Polymorphism and isomorphism in biodegradable polyesters. Progress in Polymer Science. 2009(34):605-40.

[40] Ki HC, Park OO. Synthesis, characterization and biodegradability of the biodegradable aliphatic–aromatic random copolyesters. Polymer. 2001;42:1849-61.

[41] Tokiwa Y, Suzoki T, Appl J. Hydrolysis of copolyesters containing aromatic and aliphatic ester blocks by lipase. Journal of Applied Polymer Science. 1981;26(2):441-8.

[42] Xiong X, Tam K. Hydrolytic degradation of pluronic F127/ poly (lactic acid) block copolymer nanoparticles. Macromolecules. 2004;37(34):25-30.

[43] Jin H-J, Lee B-Y, Kim M-N, Yoon J-S. Thermal and mechanical properties of mandelic acid-copolymerized poly(butylene succinate) and poly(ethylene adipate). Journal of Polymer Science, Part B: Polymer Physics. 2000;38:1504-11.

[44] Witt U, Müller R-J, Deckwer W-D. Studies on sequence distribution of aliphatic/aromatic copolyesters by high-resolution C nuclear magnetic resonance spectroscopy for evaluation of biodegradability. Makromol Chem Phys. 1996;197:1525-35.

[45] Muller RJ, Witt U, Rantze E, DeckwerWD.Architectureofbiodegradable copolyestyers containing aromatic constituents. Polymer Degradation and Stability 1998;59:203-8.

[46] Massardier-Nageotte V, Pestre C, Cruard-Pradet T, Bayard R. Aerobic and anaerobic biodegradability of polymer films and physico-chemical characterization. Polymer Degradation and Stability. 2006;91:620-7.

[47] Herrmann AS, Nickel J, Riedel U. Construction materials based upon biologically renewable resources from components to finished parts. Polym Degrad Stab. 1998;59:251-61.

[48] Artamonova SV, Demina NM. New starch based textile onlining agent for glass fibres. Fibre chemistry 1997;29:68-70.

[49] Fang Q, Hanna MA. Preparation and Characterization of Biodegradable Copolyester–Starch Based Foams. Bioresource Technology. 2001;78(2):115-22.

[50] Rhim JW, Ng PK. Natural biopolymer-based nanocomposite films for packaging applications. Critical Reviews in Food Science and Nutrition. 2007;47(4):411-33.

[51] Sekhon BS. Food nanotechnology - an overview. Nanotechnology, Science and Applications. 2010;3 1-15.

[52] Kaushik A, Arya SK, Vasudev A, Bhansali S. Recent Advances in Detection of Ochratoxin-A. Open Journal of Applied Biosensor. 2013;2:1-11.

[53] Zecheru T. Biopolymers for Military Use: Opportunities and Environment Implications - A Review. In: Elnashar M, editor. Biopolymers. Croatia: Sciyo; 2010. p. 597-612.

[54] Santos MI, Fuchs S, Gomes ME, Unger RE. Response of micro- and macrovascular endothelial cells to starch-based fiber meshes for bone tissue engineering. Biomaterials. 2007;28:240-8.

[55] Yoshimoto H, Shin YM, Terai H, Vacanti JP. A biodegradable nanofiber scaffold by electrospinning and its

potential for bone tissue engineering. Biomaterials. 2003;24:2077-82.

[56] Purna SK, Babu M. Collagen based dressings- A review. Burn. 2000;26:54-62.

[57] Qurashi MT, Blair HS, Allen SJ. Studies on modified chitosan membranes. II.Dialysis of low molecular weight metabolites. Journal of Applied Polymer Science. 1992;46:263-9.

[58] Sparke BG, Murray DG. Chitosan based wound dressing materials. US Patent, 4572906. 1986.

[59] Hoekstra A, Struszczyk H, kivekas O. Percutaneous micro-crystalline chitosan application for sealing arterial puncture sites. Biomaterials. 1998;19:1467-71.

[60] WON CS. Vascular Tissue Engineering Scaffolds from Elastomeric Biodegradable Poly(L-lactide-co-ε-caprolactone) (PLCL) via Melt Spinning and Electrospinning. USA: North Carolina State University; 2006.

[61] Wang Q, Zhang N, Hu X, Yang J, Du Y. Chitosan/starch fibers and their properties for drug controlled release. European Journal of Pharmaceutics and Biopharmaceutics. 2007;66(3):398-404.

[62] Sun YM, Hsu SC, Lai JY. Transport properties of ionic drugs in the ammonio methacrylate copolymer membranes Journal of Pharmacy Research. 2001;18(3):304-10.

[63] SA] A, H Y, B O. Membrane formation by drycast process: model validation through morphological studies. Journal of Membrane Science 2005;249(1):63-72.

[64] Younes B, Ward SC, Christie RM, Vettese S. Textile applications of commercial photochromic dyes: part 7. A statistical investigation of the influence of photochromic dyes on the mechanical properties of thermoplastic fibres. The Journal of The Textile Institute. 2019;110(5):780-90.

[65] Ajeeb F, Younes B, Khsara AK. Investigating the Relationship between Thermochromic Pigment Based knitted Fabrics Properties and Human Body Temperature. IOSR J Polym Textile Eng. 2017;4(3):44-52.

[66] Younes B. Investigating the co-effect of twist and hot drawing processes on the bio-based yarns properties. The Journal of The Textile Institute. 2020;111(2):202-13.

[67] Basel Younes SCW, Robert M Christie. Textile applications of commercial photochromic dyes: part8. A statistical investigation of the influence of photochromic dyes on thermoplastic fibres. The Journal of The Textile Institute. 2019;under

[68] Jena UCH, Elsner P. Biofunctional Textiles and the Skin. Basel (Switzerland): S. Karger; 2006.

[69] Younes B. Classification, characterization, and the production processes of biopolymers used in the textiles industry. The Journal of The Textile Institute. 2017;108(5):674-82.

[70] Aslan B, Ramaswamy S, Raina M, Gries T. Bio-composites : processing of thermoplastic biopolymers and industrial natural fibres from staple fibre blends up to fabric for composite applications. Journal of Textiles and Engineer. 1 9 9 2;19(85):47-51.

[71] Lim S, Lee J, Jang S, Lee S, Lee K, Choi H, et al. Synthetic Aliphatic Biodegradable Poly(butylene succinate)/Clay Nanocomposite Foams with High Blowing Ratio and Their Physical Characteristics. Polymer Engineering and Science. 2011;123:1316-25.

[72] Antipov EM, Dubinsky VA, Rebrov AV, Nekrasov YP, Gordeev SA,

Ungar G. Strain-induced mesophase and hard-elastic behaviour of biodegradable polyhydroxyalkanoates fibers. Polymer. 2006 47:5678-90.

[73] Lim KY, Yoon KJ, Kim BC. Highly absorbable lyocell fiber spun from celluloses/hydrolyzed starch-g-PAN solution in NMMO monohydrate. European Polymer Journal 2003;39:2115-20.

[74] Lim KY, Yoon KJ, Kim BC. Highly absorbable lyocell fiber spun from celluloses/hydrolyzed starch-g-PAN solution in NMMO monohydrate. European Polymer Journal. 2003;39:2115-20.

[75] Strnad S, Šauperl O, Fras-Zemljič L. Cellulose Fibres Functionalised by Chitosan: Characterization and Application. In: Elnashar M, editor. Biopolymers. Croatia: Sciyo; 2010. p. 181-200.

[76] Czuryszkiewicz T, Ahvenlammi J, Kortesuo P, Ahola M, Kleitz F, Jokinen M, et al. Drug release from biodegradable silica fibers. Journal of Non-Crystalline Solids. 2002;306:1-10.

[77] Kopecek J, Yang J. Hydrogels as smart biomaterials. Polymer International. 2007;56(9):1078-98.

[78] Schmack G, Jehnichen D, Vogel R, Tandler B, Beyreuther R, Jacobsen S, et al. Biodegradable fibres spun from poly(lactide) generated by reactive extrusion. Journal of Biotechnology. 2001;86:151-60.

[79] Fletcher K. Sustainable Fashion and Textiles: Design Journeys. UK: Routledge; 2012.

[80] Bond EB, Autrn JPM, Mackey LN, Noda I, Odonnell HJ, inventorsWO 02/090630 A1, Multi-component fibres comprising starch and biodegradable polymer2002.

[81] Rong H, Bhat GS. Preparation and Properties Of Cotton-Eastar Nonwovens. International Nonwovens Journal. 2003;12(2):53-7.

[82] Bhat GS, Haoming R. Effect of Binder Fibres on Processing and Properties of Thermal Bonded Cotton-Based Non-woven. International Nonwovens Technical Conference, USA. 2002.

[83] Rong H, Leon RV, Bhat GS. Statistical Analysis of the Effect of Processing Conditions on the Strength of Thermal Point-Bonded Cotton-Based Nonwovens. Textile Research Journal. 2005;75(1):35-8.

[84] Bryan BE, Marie AJP, Neil ML, Isao N, Odonnell HJ, inventorsWO 02/090629 A1, Fibres comprising starch and biodegradable polymers2002.

[85] Bhat G, Kamath MG, Mueller D, Parikh DV, McLean M. Cotton-Based Composites for Automotive Applications. Global Plastics Environmental Conference, Michigan. 2004.

[86] Twarowska-Schmidt K, Ratajska M. Biodegradability of Non-Wovens Made of Aliphatic-Aromatic Polyester. FIBRES & TEXTILES in Eastern Europe. 2005;13(1):71-4.

[87] Twarowska-Schmidt K. Evaluation of the Suitability of Some Biodegradable Polymers for the Forming of Fibres. Fibres & Textiles in Eastern Europe. 2004;12(46):15-8.

[88] Mohanlal AP. Design and Study of Biodegradable Small Diameter Woven Vascular Grafts [Master Thesis]: North Carolina State University, USA; 2003.

[89] Kwon K, Kidoaki S, Matsuda T. Electrospun nano- to microfiber fabrics made of biodegradable copolyesters:structural characteristics, mechanical properties and cell

adhesion potential. Biomaterials. 2005;26:3929-39.

[90] Kumar A, Srivastava A, Galaev IY, Mattiasson B. Smart polymers: Physical forms and bioengineering applications. Progress in Polymer Scence. 2007;32:1205-37.

[91] Lötzsch D, Ruhmann R, Seeboth A. Thermochromic Biopolymer Based on Natural Anthocyanidin Dyes Open Journal of Polymer Chemistry. 2013;3:43-7.

[92] Scott A. Spider Silk Poised For Commercial Entry. Chemical & Engineering 2014;92:24-7.

[93] Hult EL, Iotti M, Lenes M. Efficient Approach to High Barrier Packaging Using Microfibrillar Cellulose and Shellac. Cellulose, Springer. 2010;17:575-86.

[94] Knittel D, Buschmann HJ, Hipler C, Elsner P. Schollmeyer E: Funktionelle Textilien zur Hautpflege und als Therapeutikum. Akt Dermatol. 2004;30:11-7.

[95] Zemljič LF, Peršin Z, Stenius P, Stana-Kleinschek K. Carboxyl groups in pretreated regenerated cellulose fibres. Cellulose. 2008;15(5):681-90.

[96] hanmugasundaram WL. Development and characterization of cotton and organic cotton gauze fabric coated with biopolymers and antibiotic drugs for wound healing. Indian Journal of Fibre Textile Research 2012;37:146-50.

[97] Wang H-R, Chen K-M. Preparation and surface active properties of biodegradable dextrin derivative surfactants. Colloids and Surfaces A: Physicochem Eng Aspects. 2006;281:190-3.

[98] Mohanty AK, Khan MA, Hinrichsen G. Influence of chemical surface modification on the properties of biodegradable jute fabrics—polyester amide composites. Composites: Part A. 2000;31:143-50.

[99] Uretzkey G, Appelbaum AJ, Younes H. Long-term evaluation of a new selectively biodegradable vascular graft coated with polyethylene oxide-polylactic acid for right ventricle conduit. J Thorac Cardiovasc Surg. 1990;100:769-80.

[100] Madrigal-Carballoa S, Limb S, Rodrigueza G, Vilac AO, Kruegerb CG, Gunasekaranb S, et al. Biopolymer coating of soybean lecithin liposomes via layer-by-layer self-assembly as novel delivery system for ellagic acid. Journal of Functional Food. 2010;2:99-106.

[101] Pelton R. Temperature-sensitive aqueous microgels. Advances in Colloid and Interface Science. 2000;85(1):1-33.

[102] Lundin M. Adsorption of biopolymers and their layer-by-layer assemblies on hydrophilic surfaces [Doctoral Thesis]. Stockholm, Sweden: KTH Royal Institute of Technology; 2009.

[103] Jocić D, Tourrette A, Lavrič PK. Biopolymers: Biopolymer-Based Stimuli-Responsive Polymeric Systems for Functional Finishing of Textiles. Croatia: Sciyo; 2010.

[104] Lochner RH, Mater JE. Design for Quality. London Chapman and Hall; 1990.

[105] Clark KB, Wheelwright SC. The product development challenge: competing through speed, quality, and creativity. Boston: A Harvard business review book; 1995.

[106] Moreland JC, Sharp JL, Brown PJ. Lab-Scale Fiber Spinning Experimental Design Cost Comparison. Journal of Engineered Fibers and Fabrics. 2010;5(1):39-49.

[107] Ziabicki A, Jarecki L, Wasiak A. Dynamic modelling of melt spinning. Computational and Theoretical Polymer Science. 1998;8(1-2):143-57.

[108] Perera SSN. Viscoelastic Effect in the Non-Isothermal Melt Spinning Processes. Applied Mathematical Sciences. 2009;3(4):177-86.

[109] Wang GX, Matthys EF. Modelling of rapid solidification by meltspinning: effect of heat transfer in the cooling substrate. Materials Science and Engineering: A. 1991;136:85-97.

[110] Ziabicki A, Jarecki L, Wasiak A. Dynamic modelling of melt spinning. Computational and Theoretical Polymer Science. 1998;8(1/2):143-57.

[111] Kotze T. Two Dimensional Modelling of PET Melt Spinning: The effects of heat transfer limitations on the quality of PET yarn produced during melt spinning: LAP LAMBERT Academic Publishing; 2010.

[112] He W, Zhang S, Wang X. Mechanical Behavior of Irregular Fibers, Part I: Modeling the Tensile Behavior of Linear Elastic Fibers. Textile Research Journal. 2001;71(6):556-60.

[113] Bingol D, Tekin N, Alkan M. Brilliant Yellow dye adsorption onto sepiolite using a full factorial design. Applied Clay Science. 2010;50(3):315-21.

[114] Hanci TO, Alaton IA, Basar G. Multivariate analysis of anionic, cationic and nonionic textile surfactant degradation with the H 2 O 2 /UV-C process by using the capabilities of response surface methodology. Journal of Hazardous Materials. 2011;185,(1):193-203.

[115] Engin AB, Özdemir Ö, Turan M, Turan AZ. Color removal from textile dyebath effluents in a zeolite fixed bed reactor: Determination of optimum process conditions using Taguchi method. Journal of Hazardous Materials. 2008;159(2-3):348-53.

[116] Krucinska I. The influence of technological parameters on the filtration efficiency of electret needled non-woven fabrics. Article Journal of Electrostatics. 2002;56(2):143-53.

[117] Gardiner WP, Gettinby G. Experimental Design Techniques in Statistical Practice, A practical Software-Based Approach. Chichester, England: Horwood Publishing Limited; 1998.

[118] Tanguchi G. Introduction to Quality Engineering. Tokyo: Asian productivity organization; 1986

[119] Younes B. The Statistical Modelling of Production Processes of Biodegradable Aliphatic Aromatic Co-Polyester Fibres used in the Textile Industry [PhD Thesis]. UK: Heriot-Watt University; 2012.

[120] Younes B. Simple Rheological Analysis Method of Spinnable-Polymer Flow Properties Using MFI Tester. Indian Journal of Materials Science. 2015(2015):1-8.

[121] Cawse JN. Experimental Design for Combinatorial and High Throughput Materials Development. USA: John Wiley and Sons, Inc; 2003.

[122] Younes B, Fotheringham A, Mather R. Statistical Modelling of the Effect of Multi-Stage Hot Drawing on the Thermal Shrinkage and Crystallographic Order of Biodegradable Aliphatic-Aromatic Co-Polyester Fibres. Fibers and Polymers. 2011;12(6):778-88.

[123] Younes B, Fotheringham A, Dessouky HME, Haddad G. The influence of multi-stage hot-drawing on the overall orientation of biodegradable aliphatic-aromatic co-polyester fibers. Journal of Engineered Fibers and Fabrics. 2013;8(1):6-16.

[124] Younes B. A Statistical Investigation of the Influence of the Multi-Stage Hot-Drawing Process on the Mechanical Properties of Biodegradable Linear Aliphatic-Aromatic Co-Polyester Fibers. Advances in Materials Science and Applications. 2014;3(4):186-202.

[125] Brody H. Synthetic fibre materials. London: Longman group UK limited 1994.

[126] Lord P. Hand Book of Yarn Production: Technology Science and Economics England: The Textile Institute & CRC & WP 2003.

[127] Tobler-Rohr MI. Handbook of Sustainable Textile Production. UK: Woodhead Publishing Limited; 2011.

[128] Callister W. Materials Science and Engineering: An Introduction. New York: John Wiley and Sons; 1999.

[129] Chen X. Modelling and predicting textile behaviour. UK: Woodhead Publishing Ltd; 2010.

Chapter 7

Medicinal Uses with Immense Economic Potential and Nutritional Properties of *Aegle marmelos*: A Concise Review

Harekrishna Mahato and Brajesh Kumar

Abstract

Aegle marmelos is the only member of the monotype genus *Aegle* and belongs to the Rutaceae family, the citrus fruits family. This review shows the economically feasible pharmacological applications and the nutritional properties of *A. marmelos*. Each part of the *A. marmelos*/Bael tree such as root, bark, leaf, flower, fruit, and seed has therapeutic significance in Ayurvedic systems as well as other traditional medicines. In bael fruit, there are various valuable bioactive compounds that have remarkable nutritional and medicinal properties. The isolated components belong to alkaloids, terpenoids, vitamins, coumarins, tannins, carbohydrates, flavonoids, fatty acids, essential oils, and various other important biological agents. Various research studies have revealed the therapeutic effects that include antioxidant, antimicrobial, antibacterial, antidiarrheal, antiviral, antidiabetic, antiulcerative, and colitis nature of bael, which aids in inhibiting gastrointestinal problems, different cardiac issues, and other possible health benefit effects such as radioprotective effects, hepatoprotective effects, wound healing, peroxidation, inhibition of lipid, gastroprotective, cardioprotective, and free-radical scavenging (antioxidant) activity.

Keywords: *Aegle marmelos*, phytoconstituents, medicinal value, antioxidant, anticancer

1. Introduction

Food is essential for life, and the main sources of food for animals especially for humans are different parts of the plant such as leaf, stem, fruits, seeds, flowers, and roots. Plants and their parts are not only used as a food but are also used as medicines. This is due to the presence of active chemical components in parts of the plant. Nowadays, various research and study applications have focused on the active chemical constituents of the plants and their functions due to the effectiveness, less side effects, synthesized easily, and most importantly economically feasible. Traditional uses of plants and plant parts also demonstrate the direction to use plants for the specific type of disease. *A. marmelos* (L.) Correa (**Figure 1**) is a very important indigenous plant to the Indian subcontinent for over 5000 years because of its economically feasible dietery and medicinal uses. The fruits of bael are of

Figure 1.
Leaves and fruits of Aegle marmelos.

dietetic use, and the fruit pulp is used to make delights such as murabba, puddings, and juice. The subtropical fruit *A. marmelos* or commonly known as bael belongs to the Rutaceae family [1]. *A. marmelos* L., known by different names in different languages, such as Bel, Beli, and Belgiri in Hindi, Bilva, Shivadrums, Shivaphala, and Vilva in Sanskrit and Bael, Bengal quince, Indian quince, holy fruit, and golden apple in English and also known as Japanese bitter orange, stone apple, or wood apple (**Table 1**) [2]. In India, it is cultivated as a temple garden plant, and the leaves of *A. marmelos* are used to pray to Lord Shiva.

In different fields of science and technology, every part of this plant has a great utilization to cure many diseases. The variety of fruits is smaller in size than the other variety in the similar cultivated type [1]. Most common cultivar varieties of *A. marmelos*/Bael are Kagzi Gonda, Gonda no 1, Gonda no 2, Kagzi Etawah, Mirzapuri, and Baghel [1]. The potential effects of bael in different medicinal systems such as Ayurvedic, Unani, and Siddha have been already in use, and it

Languages	Different names of *A. marmelos* in corresponding languages
Hindi	Bel, Belgiri, Beli
Sanskrit	Bilva, Shivadrum, Vilva, Shivaphala
Telegu	Maredu
Kannada	Bilpatra, Malura, Kumbala,
Tamil	Kuvalum
Bengali	Bel
Gujrati	Bil
Cambodia	Phneou or Pnoi
Portugese	Marmelos
Java	Mojdo
French	Oranger du Malabar
Vietnamese	Bau Nau
Thai	Matum, Mapin

Table 1.
Various names of Aegle marmelos in different languages.

states the functionality and feasibility of bael [3]. Many researches studies on bael approved that it consists of many nutritional elements such as a large group of phytochemicals, which includes alkaloids, flavonoids, tannins, phenolic acids, and coumarins. Except the above chemicals, other class of beneficial compounds such as amino acids, vitamins, a wide range of organic acids, carbohydrates, fatty acids, fibers, minerals make bael a highly nutritious fruit with immense health benefits, and the natural availability makes it economically feasible [4]. *A. marmelos* may cure gastrointestinal diseases such as diabetes, cardiac issues, and inflammation-related problems. It can acquire protective effects against the wound, microbes, radiation, free radical generation, and depression. The natural healing power of *A. marmelos* is proved by many research studies and applications [5]. *A. marmelos* fruit is a natural source of antioxidants and functional foods. It can be used in its dehydrated form such as jam, slab, bael powder, which makes it more useful for any kinds of season. *A. marmelos* and its various products and its various uses and applications make it extremely profitable, and the availability provides greater advantage over other similar kind [6, 7]. All seeds of *A. marmelos* are surrounded by gum having adhesive, waterproofing, and oil emulsion coating properties [8]. Due to the digestive and stomachic properties of bael, it is used in the treatment of diarrhea and dysentery [9]. It acquires tannins, reducing sugar, high amount of pectin, and very small amount of nonreducing sugar [10]. Due to presoluble and insoluble mucilaginous dietary fiber and pectin, it has great nutritional values. Many minerals and vitamins such as riboflavin, ascorbic acid, thiamine, niacin, vitamin A, calcium, and phosphorus are found in bael, which makes it healthy food [11]. Focus of this review is to highlight the composition of beneficial nutrients, formulation, and their medicinal uses with availability, quantification of phytochemicals, and economic view for production. This article will also be helpful for the industries for the production of medicines, bael products, food supplements and also helpful for researchers to make it more fruitful.

2. Botanical description

The growth of the *A. marmelos* tree is slow, and the height is nearly about 762 cm and 90–120 cm in girth [12], with a somewhat spiny branches of stem are observed in the bael tree [13]. Branches are armed with straight sharp axillary thrones, 2.5 cm long. Leaves alternate, 3 – foliate, rarely 5 – foliate; petiole 2.5–6.3 cm long, terete, 2.5 cm long. Leaflets 5–10 by 2.5–6.3 cm, ovate or ovate – lanceolates, crenate, acuminate, membranous, pellucid-punctate, the lateral opposite, subsessile, the terminal long petioluled. **Table 2** shows the botanical classification of bael.

The color of the leaves is different in different phase of the leaves; it is very green and fresh to see in the early stages, but later on it looks like dark green in color [13]. On observing the bark of the tree, it has been found that it is thick in nature, and

Kingdom	**Plantae**
Order	Sapindales
Family	Rutaceae
Subfamily	Aurantioideae
Genus	Aegle
Species	*A. marmelos*

Table 2.
Botanical classification of Aegle marmelos/bael.

after slightly thrashing the bark, then gum secretion is also observed, and it becomes thicker with the presence of air [13]. *A. marmelos* flower is hypogynous and bisexual in nature, and it smells sweet. The greenish-white bael flowers are collectively held by some lateral panicles, which are held by leaf axil [1, 14]. *A. marmelos* fruit has a hard, woody exocarp, and the fruit color is mainly yellowish green, having a diameter of 5.3–7.2 cm, with an approximate weight of 77.2 g, volume of 73.7 mL, and sphericity of 93.72 ± 2.78% [15]. The color of the fruit pulp may vary from bright orange to sunset yellow, and the pulp contains seeds that are present in grooves and is surrounded by thick, clear mucilage. It contains some dots on the outer surface and also contains numerous seeds, which are hard and having thread-like hairs over their outer surface, which is white. The fruit is greenish when unripe, and upon ripening, it turns into yellowish color. Both ripe and unripe fruits are used for their medicinal values [16, 17].

3. Home remedy

Many home remedies are there, which are very useful for different diseases such as dysentery, diarrhea, digestion problem, etc. 1–3 drachms(drs) of the powder made from the mixture of *A. marmelos* fruit one part, Indian sweet fennel seeds one part, *Holarrhena antidysentrica*, two parts with Ispaghula, Sugar, Chebulic myrobalan, and plantago is very useful for chronic dysentery. There is another remedy that is very useful for both the chronic dysentery and diarrhea, and it can be prepared by mixing *Andropogon muricatus* one drop, Bael fruit four drops, *Symplocos racemosa* one drop, and *Scindapsus officinalis* one drop. In summer or very hot climate, another mixture of dried *A. marmelos* fruit pulp and Indian sweet fennel seeds both in a quantity of 21/2 drops with one drop gum of silk cotton tree, two drops honey, 1/2 drop dried ginger, and sugar is very useful in chronic diarrhea and dysentery [17].

4. Traditional uses

Extensive use of *A. marmelos* has been found in Ayurveda and traditional medicine system. In Ayurveda, the ripe fruit, which is acrid, bitter, and sweet, removes vatha and kapha, and it is good for the heart. Unani said that the ripe fruit is hot and dry; tonic, restorative, astringent, laxative; good for the heart and the brain; bad for the liver and chest [16, 18]. The unripe fruit is oily bitter, acrid, sour, tasty but difficult to digest, appetizer, binding cures dysentery, removes pain. The oil is hot and cures vatha. According to Siddha, *A. marmelos* has bitter, astringent taste, hot potency, pungent, bioavailability and reduces Vatha, Pitha, and Kapha dosha. The plant leaves can be used as an analgesic and antifungal agent [19]. Bael leaves are very useful for the treatment of asthma, hepatitis, hypogllcemia, and febrifuge. It has been also traditionally utilized for preparing medicinal mixes for numerous diseases or disorders. *A. marmelos* leaf powder may be used in the treatment of bowel syndrome [20]. It is used as a remedial measure in the treatment of beriberi [21]. *A. marmelos* leaf essential oil is reported to show various therapeutic actions [22]. The flower of *A. marmelos* has antiseptic properties and can be used as an astringent. In Thailand, the immature leaves and shoots of *A. marmelos* plant are consumed, whereas young leaves and shoots are consumed in Indonesia [23]. Flower is also used in epilepsy [24]. Marmala water, made from flower, is found helpful in catering conjunctivitis [25]. *A. marmelos* flower extract has been tested for wound-healing properties [24]. It is reported that mRNA expression has been increased by inhibition of nitric oxide, the release of prostaglandin E2 (PGE2), thus increasing wound healing in rats [26]. The roots of *A. marmelos* have been used to cure dysentery, colitis, flatulence, and fever. So, in Chyavanprash, its roots are widely used [27]. In melancholia, heart palpitation, and intermittent fever, the decoction of root and bark is used [28–30]. Root bark may be used as a fish poison and fever treatment [31]. *A. marmelos* fruit extract is very helpful in thte treatment of thyroid. During pregnancy to prevent vomiting, it is found very helpful when it is taken with rice water in boiling conditions. In abscess curing, pulp powder of unripe fruit is very helpful to cure abscess [27]. On heating, sugar is produced from starch present in unripe fruit, then the fruit extract is added with hot water and anesi and strained, the extract found is useful in dysentery. Fruit pulp in milk mixed with sugar is very helpful in the treatment of urinogenital disorders [7, 21]. *A. marmelos* fruits may be used as an antiscorbutic and stomatic agent, and it may cure diseases such as intestinal ulcer, gonorrhea, chronic constipation, and indigestion [21]. Mustard oil and powdered fruit in the ratio 2:1 are used in southern Chattisgarh in burn treatment [1]. Ripe fruit juice extract helps to cure chronic gastrointestinal disorder, piles treatment, and rectum inflammation, alongwith it reduces blood sugar levels owing to its bitter flavor [32, 33]. There is a great advantage of *A. marmelos* and its various parts that it can be preserved only by drying means; there are no chemicals are used alongwith and no costs are there to preserve for long time; really this advantage is profitable for the production of its processed food and medicines. In Thailand, it has been observed that dried *A. marmelos* leaf and fruit pulp are packed in tea bags and preserved in syrup for the further use in traditional preparation of desserts such as cake, juice, squash, nectar, jam, syrup, etc. [34]. *A. marmelos* plant acts as air purifier by absorbing harmful poisonous gas from atmosphere, making it inert/neutral and thus removing poisonous property. This is not only helps to reduce pollution, but also it clears the toxic effects and provides a fresh and healthy environment. As compared with other plants, *A. marmelos* can release higher amount of O_2 in the presence of sunlight, and plant is included in a category of "Fragrant" species because it can neutralize bad odors in air by deodorization process [32]. The unripe fruit yields yellow dye and is used for tanning calico and silk fabrics by textile industry [35]. This advantage is makes it more important and profitable for industry.

5. Nutritional properties

A. marmelos nutrients are extremely beneficial for being a healthy human, and this evidence is already proved by various research studies. Recent research studies not only focused on the nutritional benefits of *A. marmelos* but also focused on its low-cost production and marketing to build a healthy environment. Main composition of *A. marmelos* nutrients is water (60–65%), carbohydrate (9–21%), sugars (11–17%), fibers (5%), and other compositions are vitamins, minerals, glucose, various fatty acids, and amino acids [3, 36, 37]. Bael can prevents rancidity and color loss because it contains a valuable amount of vitamin A (55 mg), vitamin B and vitamin C (8 mg), which can act as an antioxidant [3, 38]. *A. marmelos* fruit pulp contains 1.7% mineral contents with copper (0.21 mg), potassium (610 mg), calcium (80 mg), phosphorous (52 mg), and iron (0.60 mg/100 g) [39]. The calorific value of *A. marmelos* (88 cal/100 g) is higher than that of apple (64 cal/100 g), guava (59 cal/100 g), and mango (36 cal/100 g) [40]. Like other fruits, *A. marmelos* is also providing different nutritional values in different stages of maturity. For estimating crude protein in dried fruits by Micro Kjeldahl's distillation, to estimate crude fat Soxhelt method is used, ash content is estimated by incineration method

[21, 41]. Xylose, threose, glucose, galactose, fructose, arabinose, sucrose, galacturonic acid are present in *A. marmelos* as a polysaccharides and as a fat present like ricinoleic, linolenic, oleic, linoleic, palmitic, and myristic [42]. The major and important parts of nutrients are found in seeds and leaf.

In *A. marmelos* seeds, there is present 34.4% oil, stearic acid 8.8%, linolenic acid 8.1%, palmitic acid 16.6%, and oleic acid 30.5% [43]. This seed oil is used in aromatherapy, cosmetics, and in compressor [44]. *A. marmelos* oil can be used in medical purposes such as an anti-inflammatory, antiseptic, antimicrobial, antioxidant, astringent, disinfectant, carminative, and cytophylactic.

6. Phytochemicals present in bael/*Aegle marmelos*

Flavonoids, carotenoids, phenolics, alkaloids, polysaccharides, terpenoids, phenylpropanoids, tannins, marmin, coumarin, etc., can found in *A. marmelos* (**Figure 2**) [45]. In leaves, the main constituents are cubebol 0.5%, γ-isogeraniol 0.2%, δ-cadinene 0.3%, γ-isogeraniol 0.2%, linalool 1.8%, menthane monoterpenoids 0.1%, α-terpineol 0.1%, α-cedrene 0.2%, germacrene B 4.7%, geraniol 0.2%, *cis*-piperitol 0.4%, A-zingiberene 0.2%, γ-isogeraniol 0.2%, α-cubebene 0.1%, α-pinene 0.2%, α-humulene 0.8%, α-copaene 0.3%, myrcene 2%, γ-elemene 0.4%, limonene 64.1%, E-phytol 0.3%, γ-curcumene 1.7%, β-funebrene 0.2%, farnesol 0.1%, E-β-ocimene 9.7%, α-farnesene 0.1%, Z-jasmone 0.1%, E-caryophyllene 2.4%, germacrene A 0.1%, trans-Linalool oxide 0.1% [46].

Many of the phytoconstituents having very good medicinal effects such as *A. marmelos* fruit pulp, fruit juice have been reported to be rich in numerous health-boosting polyphenols (**Figure 2**). Similarly, other health-benefited phytochemicals such as alkalloids, flavonoids, phenolic compounds, terpenoids have been accumulated in fruit pulp [47]. The most important polyphenols and flavonoids include alkaloids, coumarins, polysaccharides, and carotenoids. Polyphenols present in bael

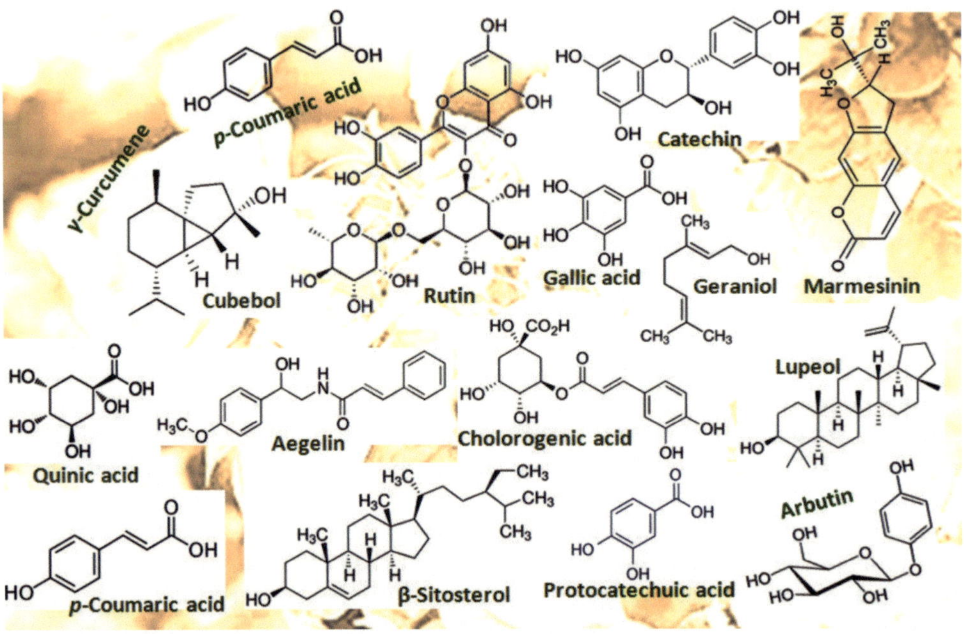

Figure 2.
List of main bioactive compounds present in Aegle marmelos.

Phytochemicals	Bael parts	Analysis method	Biological activities
Coumarins	Leaves, seed, fruit, bark, root	Colorimetric method, permanganate oxidation method	Antioxidant, antidiabetic, Anti-inflammatory, analgesic agents
Alkaloids	Leaves, fruits	Hager's test, Wagner's test, Mayer's test, Dragendorff's test	anti-inflammatory, alpha-glucosidase inhibitor, antibacterial, analgesic effects
Phenolic acids	Fruit	LC-MS, HPLC studies, Ferric chloride test	Antioxidant Activity
Terpenoid	Leaf, fruit, and bark	Noller's test	Anticancer And antimalarial
Flavonoids	Fruit	Shinoda's test	Antioxidant activity
Tannins	Unripe bael fruit	Gelatin test, Goldbeater's test	Antimicrobial, helps in reducing blood pressure
Fatty acid	Seed, fruit, leaf	Gas chromatography with a flame ionization detector	Antimicrobial activity

Table 3.
Major phytoconstituents, analysis method, and biological activities of bael plant.

depend on the maturity stage of bael [48]. From showing antioxidant activities to lipid cholesterol absorption, these phytochemicals possess immense health benefits. Caffeic cinnamamides acid, arbutin, cholorogenic acid, *p*-coumaric acid, *p*-coumaroyl, quinic acid, and protocatechuic acid such a type of phenolic compounds can be found in bael fruit juice [3]. From *A. marmelos* fruit phytochemicals such as terpenoids, flavonoids, saponins, tannins, and glycosides can be easily isolated [6]. It has been reported that compounds include γ-sitosterol, rutin, β-sitosterol, glycosides, marmeline, aegelin, marmesinin, halfordiol, phenyl ethyl, lupeol; such a type of phytochemicals can be present in a very good quantity in bael leaves [49]. Catechin, flavanols, flavones, lignin, tannins, and isoflavones fall under the polyphenol group. Lignan glucoside compounds have been isolated from the bark of *A. marmelos* tree [50]. 7, 8- dimethoxy-1- hydrox-yl 2-methyl anthraquinone and 6-hydroxy-1-methoxy-3-methyl anthraquinone have been reported to be isolated from *A. marmelos* [51], and skimmiarepin A as well as skimmiarepin C has been isolated from stem bark [52] (**Table 3**).

7. Quantitative analysis of phytochemicals

From the nutritional composition of *A. marmelos*, it is well known that the fruit is highly rich in various bioactive phytochemicals. It has been observed from various research studies that different types of phytochemicals or bioactive compounds are present in *A. marmelos*. Main class of phytochemicals are generally estimated as total phenolic count (TPC), total flavonoid content (TFC), and total carotenoids (TC). TPCs are responsible for anticancer activity of *A. marmelos*. For estimation, different processes are used, and for a particular process, there is taken a compound as a standard. For example, Folin–ciocalteu (FC) assay is used for TPC estimation with gallic acid as a standard. Similarly, TC is estimated by Gross method, and TFC estimation involves $AlCl_3$ colorimetric assay where rutin or catechin standard is used. Pearson's method is generally used for the determination of ascorbic acid, and for total alkaloids ammonium hydroxide precipitation method is used. In pulp, TPC 87.3 4 mg GAE/g dw, TFC 15.20 mg CE/g dw (catechin equivalent per gram), vitamin C 26.1 7 mg /100dw and TC 32.98 μg/g dw is present [53]. In leaf, it has been observed that there are present TFC 63.9 mg/g, total phenolic content 29.4 mg/g, alkaloids 16.0 8 mg/g, and tannins 10.5 4 mg/g [54]. In flower, TFC ethanol extract is found to be 1615.2 mg GAE/g (gallic acid equivalent/g), and aqueous extract is found to be 1496.7 mg GAE/g [55]. In bark, it has been found that total phenolic content is 14.2 mg GAE/g [56].

8. Medicinal effects of bael

8.1 Antidiarrheal activity

Figure 3a and **b** shows the biological importance of *A. marmelos* for the treatment of various diseases. The powder form of *A. marmelos* fruit is similarly effective as its fresh form. Ripe or unripe fruit is useful for the treatment diseases such as diarrhea and dysentery due to its antiprotozoal activity and also useful for treatment of constipation. In chronic dysentery, it helps to reduce bleeding, and stool loosens [57–59]. The crude aqueous extract of *A. marmelos* fruit has been reported to be nonmutagenic to *Salmonella typhimurium* strain TA 100 in the Ames assay [18]. Over 35 strains of bacteria that cause diarrhea, *Vibrio cholera*, *Escherichia coli*, and *Shigella sp* are effectively inhibited by ethanolic extract of fruit [5, 60]. Recent research studies focused on the improvement of efficiency of medicines made of *A. marmelos* for diarrhea and dysentery, and it has been observed that the chloroform extract of bael root can perform same functionality as Ciprofloxacin with the advantage of no side effects [61].

In vitro and *in vivo* studies of *A. marmelos* make it more effective and provide economic feasibility for medication. The unripe fruit pulp of *A. marmelos* shows effective potential activity over enterotoxins and also responsible for preventing the formation of gut epithelium colony [61]. Antidiarrheal activity of *A. marmelos* is not only shown by unripe fruit pulp but also by leaf, juice, and water extract of unripe fruit, and it can be observed as minimum inhibitory concentration (MIC) method [62–65]. Astringent properties of tannin present in *A. marmelos* functionalize nicely over diarrhea [66, 67]. It has been observed in Ames assay that *A. marmelos* can show nonmutagonic behavior to *S. typhimurium* strain TA100 [68, 69]. Hydroalcoholic extract of fruit is found to be nontoxic within the dose of 6 g/kg in mice [70]. It has been reported in pharmacological studies over animals that no adverse effect found after giving same dose of *A. marmelos* fruit extract for 30 days in which the maximum dose is 250 mg/kg of the body weight [53, 70, 71]. *A. marmelos* has its antigardial effect from decoction of unripe fruit [69]. Colonization studies of *E.coli* B170, *E. coli* E 134, and *Shigella flexneri* reported that colonization can be decreased by inhibiting bacterial entry [69]. The experiment of unripe fruit extracts over albino rats with inflammatory bowel disease (IBD) shows that it is very effective to reduce the intestinal inflammation. The reduction of intestinal inflammation of albino rat by the action of inflammatory mediators present in unripe *A. marmelos* fruit extract, and these are Interleukin 1, Interleukin 6, Interleukin 8, and tumor necrosis factor (TNF-α) [72]. The unripe fruit extract can also prevent intestinal mortality and secretion, which is an effective indicator of antidiarrheal action [73].

8.2 Antimicrobial activity

It has been reported that *A. marmelos* can protect against a wide variety of pathogenic organisms such as antibacterial, antitumor, antiviral, anti-inflammatory,

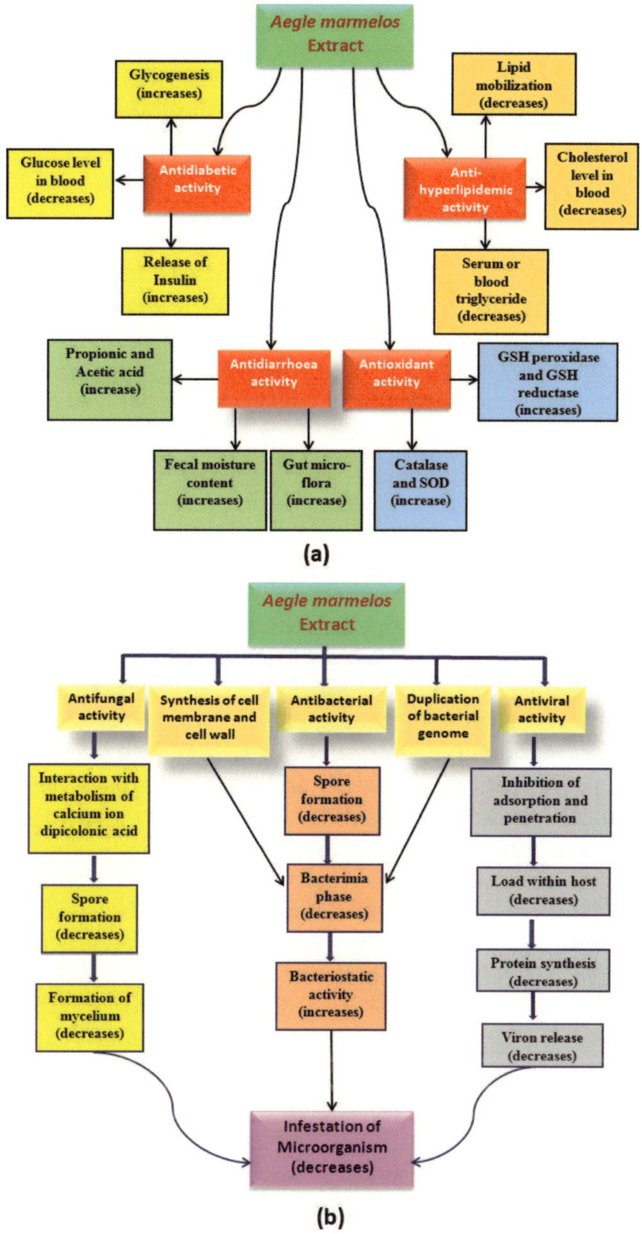

Figure 3.
Biological importance of Aegle marmelos for the treatment of various diseases and mode of action.

antifungal [1]. Marmelide, which is extracted from *A. marmelos*, has been found to have more potential and economical than ribavirin, which is an antimicrobial drug activity (**Figure 3b**). The primary phase of a replicative cycle such as adsorption as well as penetration can be inhibited by marmelide, which shows due to the virucidal activity [74]. Dhar et al. show the *A. marmelos* juice acquired antiviral potential against the Ranikhet disease virus [75]. A study of *A. marmelos* fruit extract by Rani and Khullar in 2004 has been reported that *Salmonella typhi* is well affected by methanolic extract. The antimicrobial action of essential oil extracted from *A. marmelos* not only affects *S. typhi* but also effectively reduces the spore germination assay of pathogens [76]. Different concentration of *A. marmelos* extract has different inhibition rate such as 0.05% extract dose can prevent 100% over fungi, whereas 0.04% dose can have 90% prevention and 0.03% dose having 75% prevention [77]. Antimicrobial action can also be observed in *A. marmelos* juice, fruit, and leaf extract. *A. marmelos* leaf extract is very much effective over Filariae [78] and also intestinal parasites such as *Ascaris lumbricoides* and *Entamoeba histolytica* can be prevented by fruits and its powder, which is a great advantage over other plants, and it is economically feasible [79]. *A. marmelos* seeds also take part in this action, and it has been observed in the disc diffusion assay that an anthraquinone observed in Bael seed is effective over *Candida albicans* and *Aspergillus species*. Aflatoxin is a family of toxins produced by the fungus *Aspergillus flavus*, and it has adverse effects on the plants, but it can be easily prevented by the oil of *A. marmelos* leaf extract [80]. (þ)-4-(2′-hydroxy-3′-methylbut-3′-enyloxy)-8H- [1, 3] dioxolo [4,5-h] chromen-8-one and 2-isopropenyl-4-methyl-1-oxa-cyclopenta [b] anthracene-5,10-dione are the different compounds isolated from bael and have possessed antimicrobial activity [81]. *Klebsiella pneumonia, Entercococcus faecalis, Streptococcus faecalis,* and *Micrococcus luteus* have been inhibited by *A. marmelos* fruit juice. Many times fish is attacked by the various pathogenic organisms, and this can be easily prevented by application of *A. marmelos* leaf extract. It is reported that for the survival of fish, the effectiveness of *A. marmelos* leaf extract enhanced with the dose of 5 g per Kilogram of the body weight. Leaf extract can enhance the hemoglobin level, pathogen clearance action except that it can also increase WBC count, lysosomal activity, and RBC count [81]. Dermatophytic fungi prevention is observed by methanolic *A. marmelos* extract with a concentration of 400 μg/mL [19]. *A. marmelos* extract prepared by using different solvents provides various applications with various concentrations of dose. Antiparasitic action of *A. marmelos* extract reported on cattle tick and 100% effective to prevent adult cattle tick and larvae of *Rhipicephalus microplus* with 0.3% methanolic extract. Besides this, 0.2% methanolic extract of *A. marmelos* can have 100% prevention over the larvae of *R. microplus* [82]. The concentration of 0.5–2 mg/mL of the extract of *A. marmelos* leaf is generally found very effective as an antifungal [83]. It has been reported that *A. marmelos* extracted by different methods and by using different substances such as petroleum ether, chloroform, methanol, etc., have different activities over different species [84]. *A. marmelos* extract can reduce the production rate of CD4þ T-cells CEM-green fluorescent protein in the human body affected by human immuno deficiency virus (HIV-INL4.3) [85]. *Enterococcus faecalis* can be prevented by the application of acetone extract of green fruit, which produces xanthoarnol [83]. The dose of 40 mg/mL concentration of methanolic *A. marmelos* leaf extract is applied in Agar diffusion method for both the Gram-negative and Gram-positive bacteria [86]. The adverse effect of many organisms can be prevented by a quinone compound [87], and activities of Gram-positive bacteria by alkaloid Shahidine extracted from *A. marmelos* [88]. Both the *Shigella flexneri* and *Shigella dysenteriae* can shows its resistance properties over β-lactam, and these are more convenient to β-lactam antibiotics [89]. Bark of *A. marmelos* tree shows antibacterial activity over few specific bacterial strains by Agar diffusion method where the solvents used are methanol and hexane [90]. It contains α-phellandrene and α-pinene, which acquired antimicrobial activity, and this has been observed in the estimation of MIC by Agar diffusion method [91].

8.3 Antiulcer activity

A very common disease of the gastrointestinal tract is ulcer, which can be caused by oxidative stress and presence of *Helicobacter pylori* bacteria. When mucosal flow of blood is restricted or gastro protection diminished, then the *H. pylori* bacteria affect widely as a result ulcer occurs [92]. It has been observed that *A. marmelos* seeds provide protection against aspirin-induced ulcer or pylorous-ligated ulcers on rats [93]. An *A. marmelos* seed *contains* a pyranocoumarin class compound named Luvangetin, which is responsible for antiulcer activity and very much depends on antioxidant mechanism [94]. Oral use of methanolic extract has great advantage over other medicines such as injections, and it is easier to be administered orally. In another experiment on rats that have lipopolysaccharide-induced ulcer after taking methanolic *A. marmelos* extract, improvement against ulcer is remarkable [95, 96]. It is reported that methanolic and aqueous *A. marmelos* extracts have effective antiulcer activity because they contain quercetin as antioxidant phytochemical [97].

8.4 Anticancer activity

Cancer is very big and challenging issue to survive as a healthy life in current world. The main reasons for cancer are hectic lifestyle with slow rate poisonous food. This leads to oxidative stress, which makes an easy pathway for cancer, very common disease [98]. Leaf extract of *A. marmelos* was found to be effective in inhibiting the growth of cancerous cell such as erythroleukemic HEL, T-lymphoid jurkat, melanoma Colo38, leukemic K562, breast cancer cell lines (MDA-MB 231) (isolated at M D Anderson from a pleural effusion of a patient with invasive ductal carcinoma), and Michigan Cancer Foundation-7 (MCF7), and B-lymphoid can be restricted by leaf extract of *A. marmelos* [61, 99]. Studies conducted showed that the phytoconstituents cineole, eugenol, citral, d-limonene, lupeol show antineoplastic effects. One of the important phytochemicals marmelin (1-hydroxy-5, 7-dimethoxy-2-naphtha- m-carboxaldehyde) has been found in bael, which exhibits anticancer activity against human colon cancer (HCT-116), human epithelial type 2 (HEp-2), and alveolar epithelial carcinoma cells [100]. Marmelin can also be effective to inhibit the growth of cancerous cells, leads to apoptosis, biochemical events through the induction of activation of TRADD (TNFR1-associated death domain protein), TNFα (tumor necrosis factor alpha), caspase-8, TNFRI (tumor necrosis factor receptor 1), and t-Bid, activates caspase-3 by releasing cytochrome C [101, 102]. Lupeol shows an antineoplastic effect on different neoplastic cell lines such as hepatocellular carcinoma cells, human epidermoid carcinoma cells, prostate carcinoma cell lines, human melanoma cells, and human pancreatic adenocarcinoma cells [103]. Citral is also effective to reduce the cell growth of cancerous cells by inducing apoptosis. On the other hand, Eugenol showed significant cytotoxic effects and is capable to act against malignant Caco-2 colon cells, normal human gingival fibroblast (HGF), human melanoma cell line, salivary gland tumor cell lines (HSG), and malignant HepG2 hepatoma cells [101, 102]. *A. marmelos* can show its chemopreventive action due to the presence of some important phytochemicals such as eugenol, citral, limonene, etc. [104, 105].

8.5 Anti-inflammatory activity

Inflammation is a defense mechanism and is activated when invader pathogen or foreign particles attack our system. Many indicators are there, which reflect inflammation very well; these are swelling, heat, pain, redness, etc. [67]. *A. marmelos* fruit extract shows anti-inflammatory function because of the presence of phytochemicals such as lupeol, skimmianine, and citral [106]. The main function of these phytochemicals, which are present in *A. marmelos*, is inhibiting the histamine-mediated signaling [106]. The ripened fruit juice can prevent inflammation of rectum. It has been reported that histamine-induced contraction can be antagonized by *A. marmelos* leaf and induce positive relaxant effect in the affected area [19].

8.6 Antipyretic activity

Antipyretic function in human is to lower down the body temperature. This is required only when there is fever or illness. Allopathic medicines can do that, but it has side effects, whereas *A. marmelos* plant part extracts can do the same thing without any side effects [107]. This advantage of *A. marmelos* provides a good health and lowers the expense of medicines. It has been observed in the experiment with albino rats that the temperature is decreased after applying ethanolic extract of *A. marmelos* with a dose of 200 and 400 mg per kilogram of the body weight. The body temperature reducing action of the ethanolic *A. marmelos* extract is comparable with paracetamol at a dose of 100 mg per kg of the body weight [19, 108]. Phytochemical compounds such as acetone, petroleum ether, ethanol, chloroform, diethyl ether are responsible for the antipyretic actions, and paracetamol equivalence is majorly due to the chloroform and methanol extract of *A. marmelos* [109].

8.7 Antispermatogenic activity

A. marmelos can affect the capability of fertility in male candidates. Generally fruit, leaf, and seeds can play this antifertility activity. Marmin and fagarine can reduce the fertility activity of male, and these two compounds are found in bark extract of *A. marmelos* in greater amount. It has been reported that methanolic extract of *A. marmelos* can reduce the serum testosterone level and also reduce the weight of the reproductive organ [61]. This reduction depends on the concentration of the dose taken and time duration [61]. Sperm viability, density, motility, and sperm acrosomal integrity all these can reduce by the effect of methanolic extract of *A. marmelos*. Condensation and degeneration of nuclear chromatin and exfoliation of extended spermatids were found in the study of histopathology and the availability of space within germinal epithelium denoting necrosis and testicular cytotoxicity. The experiment on rats has been observed that *A. marmelos* extract affects infertility, but this is temporary and totally time- and dose-dependent, and the most important thing is after the removal of the treatment, there is a complete restoration of morphological and physiological parameters [110]. It opens a new possibility to the development of herbal contraceptive for males because it can stop fertile power, and after the removal of treatment, it can do complete restoration of the fertile power [111].

8.8 Diuretic activity

Through urination, our body expels out the salt and other materials with water and balances body temperature and water level. Those suffering from water retention and anxiety need medication. *A. marmelos* leaves and roots are natural source of providing diuretic activity to increase urination level for such individuals, whereas roots are more beneficial than leaves. At higher dose (500 mg/kg) ethanolic extracts of *A. marmelos* fruit result in significant increase in sodium excretion. Extracted fraction of ethyl acetate, petroleum ether, and chloroform of *A. marmelos* fruit showed effective diuretic activity [61].

8.9 Antigenotoxic activity

Antigenotoxic activity of *A. marmelos* fruit extract can be estimated by using the peripheral human blood lymphocytes and *E. coli* PQ37. Chromotest (*E. coli* PQ37) SOS response is induced by aflatoxin B1 and H_2O_2, and it is prevented by the acetone and methanol. In general, SOS response is a response to DNA damage, in which cell cycle is arrested followed by DNA repair and mutagenesis is induced. Genotoxicity of aflatoxin B1 and hydrogen peroxide is inhibited by 84.65 and 70.48%, respectively, by the action of methanol extract of *A. marmelos* fruit [19]. Presence of different polyphenolic components in the fruit extract accounts for the antigenotoxic activity and has the ability to save DNA from the S9-dependent mutagens and reactive oxygen species. Several researchers revealed that in metabolism of aflatoxin B1, the enzyme activity can be inhibited by the polyphenols present in *A. marmelos* fruit extract [112].

8.10 Radioprotective activity

Radiotherapy is used in the treatment of cancer, but it has some side effects, which are harmful. The side effect occurs due to invariable exposure of radio sensitivity to the cytotoxic effects of ionized radiation on treatment. At the time of ionized radiation treatment, many cells are damaged because of the formation of free radicals, which can be effectively countered by antioxidant compounds [113]. It has been reported that bael fruit juice has potential to minimize the effect of free radicals; as a result, the severity reduces and survival probability is enhanced in mice [114]. In the radiation treatment of mouse in intestine and bone marrow, the detrimental effect of radiation is significantly reduced by *A. marmelos* fruit extract [115]. The hydroalcoholic extract of *A. marmelos* when administered intraperitoneally in mice has been watched that there is increase in survival rate after exposure to gamma radiation at a lethal dose of 10 Gy [116]. The inhibition power to the radiation of *A. marmelos* fruit and leaf extracts is due to free radical scavenging and lipid peroxidation associated by the improvement in GSH concentration in the spleen, kidney, intestine, and liver [115, 117]. On comparison of inhibition power to the radiation between the *A. marmelos* plant parts, then the leaf extract was found better, and this can provide extra protection from hematopoietic and gastrointestinal damages.

8.11 Antidiabetic activity

Nowadays, one of the very common diseases is diabetes mellitus. Stressful lifestyle, fastfood consumption, and genetic issues make diabetes a very common disease [118]. Specially, oxidative stress of the body is closely related with this disease. *A. marmelos* fruit juice increases the utilization of glucose by stimulation and the effect looks like similar to the function of insulin [119, 120]. Fruit juice can improve blood circulation by dilating the blood vessels; as a result of this, blood clot formation is reduced, which helps to remove diabetes (**Figure 3a**). Amino acids and dietary fiber are present in the fruit juice of *A. marmelos* and can absorb moderate sugar and promote insulin sensitivity [40]. *A. marmelos* fruit extract shows multifunctional activity over diabetes, it is not only reduces the glucose level in blood and glycosylated hemoglobin but also simultaneously enhances the plasma insulin as well as glycogen in diabetic rats [121]. It has been reported that methanolic extracts of *A. marmelos* leaf reduce the level of glucose in blood of Wistar albino rats according to the doses applied; as it increases, the serum insulin and so the level of glucose automatically decrease [122, 123]. Insulin secretion can be affected by coumarin, which is present in *A. marmelos* fruit juice [124]. Both the ways of administration means oral and intraperitoneal exhibit hypoglycemic effect. The main mechanism of action of *A. marmelos* is improvement of insulin secretion or stimulation of glucose or both.

8.12 Wound-healing activity

Wound healing is a multistep process and involves different techniques such as contraction of collagen lattice production, cell proliferation, inflammation, etc. Pain is the common symptoms of wound, and main key factor for wound healing is the discharge of reactive oxygen species (ROS), which helps in healing [125]. In an experiment, reports on the Wistar albino rats showed that methanolic *A. marmelos* seed extract ointment can recover the wound perfectly [126]. The functionality of *A. marmelos* on wound healing is like that of nitrofurazone drug [127]. Healing process is boosted when myeloperoxidase (MPO) level and free radicals touch the minimum level. On the other hand, if antioxidant activity increases, then the rate of healing mechanism is also increased. The phytochemicals present in *A. marmelos* are very much effective for wound healing because they can increase the antioxidant activity [24].

8.13 Immunomodulatory effect

A. marmelos can improve the immunity upto a particular stage depending on the situation. It has been observed that after oral administration of different doses of *A. marmelos,* both the neutrophil adhesion and phytogenic index are increased. Bovine pasteurella multocida can reduce the mortality rate in mice at both low and high doses. Administrations of lower doses are much more effective than higher doses [128]. Another study on fish reported that survival chances are more with *A. marmelos* leaf administration [129]. Bael can increase enzyme activity, RBC count, and WBC count, which can lead to a healthy system; as a result, immunity is increased.

8.14 Nephroprotective activity

A. marmelos leaf extract shows its nephroprotective function. Nephrotoxicity is mainly observed due to the enhancement of blood urea nitrogen, serum creatinine, and MDA level. The study of gentamicin nephrotoxicity reported that *A. marmelos* leaf extract can dominantly reduce the enhanced level of MDA, blood urea nitrogen, and serum creatinine [130]. The cisplatin-promoted nephrotoxicity can also be protected by *A. marmelos* leaf extract [131].

8.15 Antiasthmatic activity

In phelgm and asthma, decoction of *A. marmelos* leaf can reduce these effectively. Guinea pig ileum study reveals that antagonistic effect can be shown by tracheal chain on contraction produced by histamine [132, 133]. Histamine release from mast cells can be prevented by aegeline situated in *A. marmelos* leaf [134]. Histamine release is the main reason for asthma and phelgm. *A. marmelos* leaves effectively reduce the release of histamine; that is why it is very helpful for controlling asthma.

8.16 Antithyroid activity

A. marmelos leaf extract can reduce the enhanced level of thyroid hormones in serum. This function of *A. marmelos* leaf is due to the presence of scopoletin in it, which is actually responsible for the reduction of thyroid hormones in serum. The activity of Scopoletin is as close to the activity of Propylthiouracil [135]. The dose is experimented on thyroxine-treated animals and found that Scopoletin 1 mg /Kg for sevèn days is most effective.

9. Conclusion and future prospectives

In recent years, scientific studies have validated many of the ethnomedicinal uses and reports on various parts of *A. marmelos. A. marmelos* has immense medicinal actions over animals, and it may replace many medicines of allopathy. The functionalities of many allopathic medicines can be done by different parts and its extracts of *A. marmelos*. As a result, this not only reduces the expense of medicines but also reduces the chances of new other diseases as it has no side effects. This is a great advantage over allopathy medicines, and simultaneously it is economically feasible. The phytochemicals present in *A. marmelos* are very effective, and these are responsible for the various therapeutic activities, and this is the reason for more investigation and research required to know the functionalities of phytoconstituents at molecular level, which can provide more efficient medicine. Planting this tree is very easy and simple; therefore, the availability is fruitful, which is a great advantage. Each and every part of this tree can be used as a medicine, so all the parts are in use, which reflects more marketing from one tree; it enhances the productivity, and economically there is a profitable position for the manufacturer industry. This review article tried to focus on the medicinal uses with economically feasible production and marketing and found exceptionally well, although more research required for the improvement of medicinal functionalities and productivity.

Acknowledgements

This scientific work has been funded by the TATA College, Kolhan University, Chaibasa, India.

Conflicts of interest

The authors declare that there is no conflict of interests regarding the publication of this article.

Author details

Harekrishna Mahato and Brajesh Kumar*
Department of Chemistry, TATA College, Kolhan University, Chaibasa, Jharkhand, India

*Address all correspondence to: krmbraj@gmail.com

IntechOpen

© 2022 The Author(s). Licensee IntechOpen. This chapter is distributed under the terms of the Creative Commons Attribution License (http://creativecommons.org/licenses/by/3.0), which permits unrestricted use, distribution, and reproduction in any medium, provided the original work is properly cited.

References

[1] Sharma PC, Bhatia V, Bansal N, Sharma A. A review on Bael tree. Indian Journal of Natural Products and Resources. 2007;**6**:171-178

[2] Lim TK. Edible Medicinal and Non Medicinal Plants. Netherlands: Springer; 2015. DOI: 10.1007/978-94-017-9511-1

[3] Bhardwaj RL, Nandal U. Nutritional and therapeutic potential of Bael (*Aegle marmelos* Corr.) fruit juice: A review. Nutrition & Food Science. 2015;**45**:895-919

[4] Bhardwaj RL. Role of Bael fruit juice in nutritional security of Sirohi tribals. Benchmark Survey Report of Sirohi Tribals, Krishi Vigyan Kendra, AU, Jodhpur 2014:11-37

[5] Manandhar B, Paudel KR, Sharma B, Karki R. Phytochemical profile and pharmacological activity of *Aegle marmelos* Linn. Journal of Integrative Medicine. 2018;**16**:153-163

[6] Rajan S, Gokila M, Jency P, Brindha P, Sujatha RK. Antioxidant and phytochemical properties of *Aegle marmelos* fruit pulp. International Journal of Current Pharmaceutical Research. 2011;**3**:65-70

[7] Anurag S, Pragati KSHK, Ashutosh U. Bael (*Aegle marmelos* Correa) products processing: A review. African Journal of Food Science. 2014;**8**:204-215

[8] Jindal M, Kumar V, RanaV TAK. *Aegle marmelos* fruit pectin for food and pharmaceuticals: Physico-chemical, rheological and functional performance. Carbohydrate Polymers. 2013;**93**(2): 386-394

[9] Baliga MS, Bhat HP, JosephN FF. Phytochemistry and medicinal uses of the Bael fruit (*Aegle marmelos* Correa): A concise review. Food Research International. 2011;**44**(7):1768-1775

[10] Chakraborty K, Saha J, Raychaudhuri U, Chakraborty R. Optimization of bioprocessing parameters using response surface methodology for bael (*Aegle marmelos* L.) wine with the analysis of antioxidant potential, colour and heavy metal concentration. Nutrafoods. 2015;**14**(1):39-49

[11] Bhattacherjee AK, Dikshit A, Pandey D, Tandon DK. High performance liquid chromatographic determination of marmelosin and psoralen in bael (*Aegle marmelos* (L.) Correa) fruit. Journal of Food Science and Technology. 2015;**52**(1):597-600

[12] Jagtap AG, Shirke SS, Phadke AS. Effect of a polyherbal formulation on experimental models of inflammatory bowel diseases. Journal of Ethnopharmacology. 2004;**90**:195-204

[13] Bhar K, Mondal S, Suresh P. An eye-catching review of *Aegle marmelos* L. (golden apple). Pharmacognosy Journal. 2019;**11**:207-224

[14] Patel AR, Garach D, Chakraborty M, Kamath J. *Aegle marmelos* (Linn.): A therapeutic boon for human health. International Journal of Research in Ayurveda and Pharmacy. 2012;**3**(2): 159-163

[15] Sonawane A, Pathak SS, Pradhan RC. Physical, thermal, and mechanical properties of Bael fruit. Journal of Food Process Engineering. 2020:**43**;e13393

[16] Kirtikar KR, Basu BD. Indian Medicinal Plants. 2nd ed. Allahabad, India: Lalit Mohan Basu; 1939. pp. 499-502

[17] Nadkarni MK. Indian Materia Medica. 3rd ed. India: Popular Prakashan; 2009. pp. 45-49

[18] Daniel M. Medicinal plants-chemistry and properties of medicinal plants. IBH Publication. 2006;**147**:6

[19] Balakumar S, Rajan S, Thirunalasundari T, Jeeva S. Antifungal activity of *Aegle marmelos* (L.) Correa (Rutaceae) leaf extract on dermatophytes. Asian Pacific Journal of Tropical Biomedicine. 2011;**1**(4):309-312

[20] Atul NP, Nilesh VD, Akkatai AR, Kamlakar SK. A review on *Aegle marmelos*: A potential medicinal tree. International Research Journal of Pharmacy. 2012;**3**:86-91

[21] Sampath KP, Umadevi M, Bhowmik D, Singh DM, Dutta AS. Recent trends in medicinal uses and health benefits of Indian traditional herbs *Aegle marmelos*. Pharma Innovation. 2012;**1**:57-65

[22] Fawzi Mahomoodally M, Mollica A, Stefanucci A, Zakariyyah Aumeeruddy M, Poorneeka R, Zengin G. Volatile components, pharmacological profile, and computational studies of essential oil from *Aegle marmelos* (Bael) leaves: A functional approach. Industrial Crops and Products. 2018;**126**:13-21

[23] Lakht-e-Zehra A, Dar NG, Saleem N, Soomro UA, Afzal W, Naqvi B, et al. Nutritional exploration of leaves, seed and fruit of bael (*Aegle marmelos* L.) grown in Karachi region. Pakistan Journal of Biochemistry & Molecular Biology. 2015;**48**(3):61-65

[24] Gautam MK, Purohit V, Agarwal M, Singh A, Goel RK. *In vivo* healing potential of *Aegle marmelos* in excision, incision, and dead space wound models. Scientific World Journal. 2014;**9**:459-486

[25] Mani A, Singh A, Jain N, Misra S. Flowering, Fruiting and physio-chemical characteristics of bael (*Aegle marmelos* correa.) grown in northern districts of West Bengal. Current Journal of Applied Science and Technology. 2017;**23**:1-8

[26] Shirwaikar A, Shenoy R, Udupa AL, Udupa SL, Shetty S. Wound healing property of ethanolic extract of leaves of *Hyptis suaveolens* with supportive role of antioxidant enzymes. Indian Journal of Experimental Biology. 2003;**41**:238-241

[27] Kala CP. Ethnobotany and Ethnoconservation of *Aegle marmelos* (L.) Correa. Indian Journal of Traditional Knowledge. 2006;**5**(4):537-540

[28] Sharma J, Painuli RM, Gaur RD. Plants used by the rural communities of district Shahjahanpur, Uttar Pradesh. Indian Journal of Traditional Knowledge. 2010;**9**:798-803

[29] Yadav KP, Gupta LN, Shukla AK. Outcome of bilwa (*Aegle marmelos*) phala majja churna on intestinal transit. Ayurpharm International Journal of Ayurveda and Allied Sciences. 2015;**4**:23-28

[30] Kakiuchi N, Senaratne LRE, Huang SL, Yang XW, Hattori M, Pilapitiya U, et al. Effects of constituents of Beli (*Aegle marmelos*) on spontaneous beating and calcium-paradox of myocardial cells. Planta Medica. 1991;**57**:43-47

[31] Lock JM, Gaur RD. Flora of the district garhwal north west himalaya (with ethnobotanical notes). Kew Bulletin. 2001;**56**:251

[32] Dhankhar S, Ruhil S, Balhara M, Dhankhar S, Chhillar AK. *Aegle marmelos* (Linn.) Correa: A potential source of phytomedicine. Journal of Medicinal Plants Research. 2011;**5**(9):1497-1507

[33] Kirtikar BD, Basu KR. Indian medical plants. Goyal Offset Printers. 1984;**3**:2143-2155

[34] Chauhan H, Kaul RK, Ahmed N, Gupta P, Anjum A. Preparation and evaluation of Bael (*Aegle marmelos*) vermouth. International Journal of Food and Fermentation Technology. 2016;**6**(1):41-47

[35] Mali SS, Dhumal RL, Havaldar VD, Shinde SS, Jadhav NY, Gaikwad BS. A systematic review on *Aegle marmelos* (Bael). Research Journal of Pharmacognosy and Phytochemistry. 2020;**12**(1):31-36

[36] Gupta D, John P, Kumar P, Jain J. Evaluation of antioxidant activity of unripe *Aegle marmelos* corr. Fruits. Journal of Applied Pharmaceutical Sciences and Research. 2018;**1**(1):4-7

[37] Nandal UA. Handbook of Foods and Nutritional Biochemistry: A Complete Source Book. Agrobios India; 2013

[38] Murthy HN, BhatMA DD. Bioactive compounds of Bael *(Aegle marmelos* (L.) Correa). In: Bioactive Compounds in Underutilized Fruits and Nuts. New York: Springer; 2020. pp. 459-486

[39] Singh K, Chakraborty I, Chaurasiya AK. Bael preserve syrup as booster of human health as a health drink. The Bioscan. 2014;**9**(2):565-569

[40] Bhardwaj RL, Nandal U, Pal A, Jain S. Bioactive compounds and medicinal properties of fruit juices. Fruits. 2014;**69**(5):391-412

[41] Sharma K, Chauhan ES. Nutritional and phytochemical evaluation of fruit pulp, powder of *Aegle marmelos* (Bael). JCPS. 2017;**10**(2):809-814

[42] Bhattacherjee AK, Dikshit A, Tandon DK. Nutraceutical changes during ripening of bael [*Aegle marmelos* (L.) Correa] fruits harvested at different maturity periods. Indian Journal of Traditional Knowledge. 2020;**19**(2): 416-422

[43] Kumar KS, Umadevi M, Bhowmik D, Singh DM, Dutta AS. Recent trends in medicinal uses and health benefits of Indian traditional herbs *Aegle marmelos*. The Pharma Innovation. 2012;**1**(4):57-65

[44] Bajaniya VK, Kandoliya UK, Bodar NH, Bhadja NV, Golakiya BA. Fatty acid profile and phytochemical characterization of bael seed (*Aegle marmelos* L.) oil. International Journal of Current Microbiology and Applied Sciences. 2015;**4**(2):97-102

[45] Chhikara N, Abdulahi B, Munezero C, Kaur R, Singh G, Panghal A. Exploring the nutritional and phytochemical potential of sorghum in food processing for food security. Nutrition & Food Science. 2019;**49**(2):318-332

[46] Yadav SD, Muley RD, Kulkarni AU, Sarode BN. Study of Bael oil (*Aegle marmelos*) and its properties. Women. 2013;**1**:1904-1913

[47] Snehlata K, Sheel R, Kumar B. Evaluation of phytochemicals in polar and non polar solvent extracts of leaves of *Aegle marmelos* (L.). Journal of Biotechnology and Biochemistry. 2018;**4**(5):31-38

[48] Gurjar PS, Bhattacherjee AK, Singh A, Dikshit A, Singh VK. Characterization of nutraceuticals in bael powder prepared from fruits harvested at different developmental stages. Indian Journal of Traditional Knowledge. 2019;**18**:724-730

[49] Bisht NV, Johar V. Bael (*Aegle marmelos*) extraordinary species of India: A review. International Journal of Current Microbiology and Applied Sciences. 2017;**6**:1870-1887

[50] Ohashi K, Watanabe H, Okumura Y, Uji T, Kitagawa I. Indonesian medicinal plants. XII. Four isomeric lignanglucosides from the bark of Aegle marmelos (Rutaceae). Chemical & Pharmaceutical Bulletin. 1994;**42**(9): 1924-1926

[51] Srivastava SD, Srivastava S, Srivastava SK. New anthraquinones

from the heartwood of *Aegle marmelos*. Fitoterapia. 1996;**67**:83-84

[52] Radhika Samarasekera JKR, Khambay BPS, Hemalal KP. A new insecticidal protolimonoid from *Aegle marmelos*. Natural Product Research. 2004;**18**:117-122

[53] Kamalakkannan N, Prince PSM. Effect of *Aegle marmelos* Correa. (Bael) fruit extract on tissue antioxidants in streptozotocin diabetic rats. Indian Journal of Experimental Biology. 2003;**41**:1285-1288

[54] Karunanayake EH, Welihinda J, Sirimanne SR, Adorai GS. Oral hypoglycaemic activity of some medicinal plants of Sri Lanka. Journal of Ethnopharmacology. 1984;**11**:223-231

[55] Sabu MC, Kuttan R. Antidiabetic activity of *Aegle marmelos* and its relationship with its antioxidant properties. Indian Journal of Physiology and Pharmacology. 2004;**48**(1):81-88

[56] Benni JM, Jayanthi MK, Suresha RN. Evaluation of the anti-inflammatory activity of *Aegle marmelos* (Bilwa) root. Indian Journal of Pharmacology. 2011;**43**:393-397

[57] Ashkenazi S, Pickering LK. Pathogenesis and diagnosis of bacterial diarrhoea. European Journal of Clinical Microbiology & Infectious Diseases. 1989;**8**:203-206

[58] Gupta D, John PP, Pankaj K, Kaushik R, Yadav R. Pharmacological review of *Aegle marmelos* corr. Fruits. IJPSR. 2011;**2**(8):2031-2036

[59] Wallace JL, Granger DN. The cellular and molecular basis of gastric mucosal defense. The FASEB Journal. 1996;**10**:731

[60] Mazumder R, Bhattacharya S, Mazumder A, Pattnaik AK, Tiwary PM, Chaudhary S. Antidiarrhoeal evaluation of *Aegle marmelos* (correa) Linn. Root extract. Phytherapy Research. 2006;**20**:82-84

[61] Rahman S, Parvin R. Therapeutic potential of *Aegle marmelos* (L.)-an overview. Asian Pacific Journal of Tropical Disease. 2014;**4**:71-77

[62] Dhuley JN. Investigation on the gastroprotective and antidiarrhoeal properties of *Aegle marmelos* unripe fruit extract. Hindustan Antibiotics Bulletin. 2003;**45-46**:41-46

[63] Sondhi N, Bhardwaj R, Kaur S, Kumar N, Singh B. Isolation of 24-epibrassinolide from leaves of *Aegle marmelos* and evaluation of its antigenotoxicity employing *Allium cepa* chromosomal aberration assay. Plant Growth Regulation. 2008;**54**:217-224

[64] Sekar DK, Kumar G, Karthik L, Rao KVB. A review on pharmacological and phytochemical properties of *Aegle marmelos* (L.) Corr. Serr. (Rutaceae). Asian Journal of Plant Science and Research. 2011;**1**(2):8-17

[65] Joshi PV, Patil RH, Maheshwari VL. *In vitro* antidiarrhoeal activity and toxicity profile of *Aegle marmelos* Correa ex Roxb. Dried fruit pulp. Natural Product Radiance. 2009;**8**(5):498-502

[66] Shoba FG, Thomas M. Study of antidiarrhoeal activity of four medicinal plants in castor-oil induced diarrhoea. Journal of Ethnopharmacology. 2001;**76**:73-76

[67] Maity P, Hansda D, Bandyopadhyay U, Mishra DK. Biological activities of crude extracts and chemical constituents of bael, *Aegle marmelos* (L.) Corr. Indian Journal of Experimental Biology. 2009;**47**:849-861

[68] Daniel M. Medicinal Plants: Chemistry and Properties. London: CRC Press; 2006

[69] Rakulini R, Kalaichelvi S, Prasad S. A review of anti-diarrheal activity of *Aegle marmelos*. Journal of Complementary and Alternative Medical Research. 2019;7:1-10

[70] Jagetia GC, Venkatesh P, Baliga MS. Fruit extract of *Aegle marmelos* protects mice against radiation-induced lethality. Integrative Cancer Therapies. 2004;**3**:323-332

[71] Kamalakkannan N, Prince PS. The effect of *Aegle marmelos* fruit extract in streptozotocin diabetes: A histopathological study. Journal of Herbal Pharmacotherapy. 2005;**5**:87-96

[72] Jagtap AG, Shirke SS, Phadke AS. Effect of polyherbal formulation on experimental models of inflammatory bowel diseases. Journal of Ethnopharmacology. 2004;**90**:195-204

[73] Liu ESL, Cho CH. Relationship between ethanol-induced gastritis and gastric ulcer formation in rats. Digestion. 2000;**62**:232-239

[74] Badam L, Bedekar SS, Sonawane KB, Joshi SP. *In vitro* antiviral activity of Bael (*Aegle marmelos* Corr) upon human coxsackieviruses B1-B6. Journal of Communication Disorders. 2002;**34**:88-89

[75] Dhar ML, Dhar MM, Dhawan BN, Mehrotra BN, Ray C. Screening of Indian plants for biological activity: I. Indian Journal of Experimental Biology. 1968;**6**:232-247

[76] Rani P, Khullar N. Antimicrobial evaluation of some medicinal plants for their anti-enteric potential against multi-drug resistant salmonella typhi. Phytherapy Research. 2004;**18**:670-673

[77] Rana BK, Singh UP, Taneja V. Antifungal activity and kinetics of inhibition by essential oil isolated from leaves of *Aegle marmelos*. Journal of Ethnopharmacology. 1997;**57**:29-34

[78] Sahare KN, Anandhraman V, Meshram VG, Meshram SU, Reddy MVR, Tumane PM, et al. Anti-microfilarial activity of methanolic extract of *Vitex negundo* and *Aegle marmelos* and their phytochemical analysis. Indian Journal of Experimental Biology. 2008;**46**:128-131

[79] Walker AF. Herbs of choice: The therapeutic use of phytomedicinals. Trends in Pharmacological Sciences. 1995;**16**:76-77

[80] Singh P, Kumar A, Dubey NK, Gupta R. Essential oil of *Aegle marmelos* as a safe plant-based antimicrobial against postharvest microbial infestations and aflatoxin contamination of food commodities. Journal of Food Science. 2009;**74**:M302-M307

[81] Pratheepa V, Ramesh S, Sukumaran N. Immunomodulatory effect of *Aegle marmelos* leaf extract on freshwater fish *Cyprinus carpio* infected by bacterial pathogen *Aeromonas hydrophila*. Pharmaceutical Biology. 2010;**48**:1224-1239

[82] Elango G, Rahuman AA, Bagavan A, Kamaraj C, Zahir AA, Rajakumar G, et al. Efficacy of botanical extracts against Japanese encephalitis vector, *Culex tritaeniorhynchus*. Parasitology Research. 2010;**106**:481-492

[83] Chakthong S, Weaaryee P, Puangphet P, Mahabusarakam W, Plodpai P, Voravuthikunchai SP, et al. Alkaloid and coumarins from the green fruits of *Aegle marmelos*. Phytochemistry. 2012;**75**:108-113

[84] Kothari S, Minda M, Tonpay SD. Anxiolytic and antidepressant activities of methanol extract of *Aegle marmelos* leaves in mice. Indian Journal of Physiology and Pharmacology. 2010;**54**:318-328

[85] Sabde S, Bodiwala HS, Karmase A, Deshpande PJ, Kaur A, Ahmed N, et al.

Anti-HIV activity of Indian medicinal plants. Journal of Natural Medicines. 2011;**65**:662-669

[86] Mujeeb F, Bajpai P, Pathak N. Phytochemical evaluation, antimicrobial activity, and determination of bioactive components from leaves of *Aegle marmelos*. BioMed Research International. 2014;**2014**:497606. DOI: 10.1155/2014/497606

[87] Rejiniemon TS, Arasu MV, Duraipandiyan V, Ponmurugan K, Al-Dhabi NA, Arokiyaraj S, et al. *In-vitro* antimicrobial, antibiofilm, cytotoxic, antifeedant and larvicidal properties of novel quinone isolated from *Aegle marmelos* (Linn.) Correa. Annals of Clinical Microbiology and Antimicrobials. 2014;**13**:48. DOI: 10.1186/s12941-014-0048-y

[88] Jamal MAHM, Rahman MS, Hossain MB, Sharma SP, Chung HJ, Kim HJ, et al. Antibacterial properties and chemical composition of essential oil from *Aegle marmelos* (L.) corr. Leaves growing in Bangladesh. Journal of Essential Oil Bearing Plants. 2017;**20**:155-174

[89] Raja SB, Murali MR, Devaraj SN. Differential expression of ompC and ompF in multidrug-resistant *Shigella dysenteriae* and *Shigella flexneri* by aqueous extract of *Aegle marmelos*, altering its susceptibility toward β-lactam antibiotics. Diagnostic Microbiology and Infectious Disease. 2008;**61**:321-328

[90] Nemkul CM, Bajracharya GB, Shrestha I, Gan C, Bajracharya B. Phytochemical, antibacterial and DPPH free radical scavenging evaluations of the barks of *Aegle marmelos* (L.) Correa. Journal of Pharmacognosy and Phytochemistry. 2018;**7**:1637-1641

[91] Ibrahim NA, El-Sakhawy FS, Mohammed MMD, Farid MA, Abdel-Wahed NAM, Deabes DAH. Chemical composition, antimicrobial and antifungal activities of essential oils of the leaves of *Aegle marmelos* (L.) Correa growing in Egypt. Journal of Applied Pharmaceutical Science. 2015;**5**(2):1-5

[92] Biswas K, Bandyopadhyay U, Chattopadhyay I, Varadaraj A, Ali E, Banerjee RK. A novel antioxidant and antiapoptotic role of omeprazole to block gastric ulcer through scavenging of hydroxyl radical. The Journal of Biological Chemistry. 2003;**278**: 10993-11001

[93] Goel RK, Maiti RN, Manickam M, Ray AB. Antiulcer activity of naturally occurring pyrano-coumarin and isocoumarins and their effect on prostanoid synthesis using human colonic mucosa. Indian Journal of Experimental Biology. 1997;**35**:1080-1083

[94] Madhu C, Hindu K, Sudeepthi CD, Maneela P, Venkat Reddy K, Bhagya SB. Anti ulcer activity of aqueous extract of *Aegle marmelos* leaves on rats. Asian Journal of Pharmaceutical Research. 2012;**2**:132-135

[95] Ramakrishna YG, Savithri K, Kist M, Devaraj SN. *Aegle marmelos* fruit extract attenuates *helicobacter pylori* lipopolysaccharide induced oxidative stress in Sprague Dawley rats. BMC Complementary and Alternative Medicine. 2015;**15**:375. DOI: 10.1186/s12906-015-0915-x

[96] Das S, Roy C. The protective role of *Aegle marmelos* on aspirin-induced gastro-duodenal ulceration in albino rat model: A possible involvement of antioxidants. Saudi Journal of Gastroenterology. 2012;**18**:188-194

[97] Sharma AK, Bharti S, Goyal S, Arora S, Nepal S, Kishore K, et al. Upregulation of PPARγ by *Aegle marmelos* ameliorates insulin resistance and β-cell dysfunction in high fat diet fed-streptozotocin induced type 2

diabetic rats. Phytherapy Research. 2011;**25**:1457-1465

[98] Sardana RK, Chhikara N, Tanwar B, Panghal A. Dietary impact on esophageal cancer in humans: A review. Food & Function. 2018;**9**(4): 1967-1977

[99] Lampronti I, Martello D, Bianchi N, Borgatti M, Lambertini E, Piva R, et al. *In vitro* antiproliferative effects onhuman tumor cell lines of extracts from the Bangladeshi medicinal plant Aegle marmelos Correa. Phytomedicine. 2003;**10**(4):300-308

[100] Murthy HN, Bhat MA, Dalawai D. Bioactive Compounds of Bael (*Aegle marmelos* (L.) Correa). In Bioactive Compounds in under-Utilized Fruits and Nuts. New York, NY: Springer; 2020. pp. 459-486

[101] Perumal A, Krishna S, Madhusree M. GC-MS analysis, antioxidant and antibacterial activities of ethanol extract of leaves of *Aegle marmelos* (L.) corraša. Journal of Drug Delivery and Therapeutics. 2018;**8**(4):247-255

[102] Subramaniam D, Giridharan P, Murmu N, Shankaranarayanan NP, May R, Houchen CW, et al. Activation of apoptosis by 1-hydroxy-5, 7-dimethoxy-2-naphthalene-carboxaldehyde, a novel compound from *Aegle marmelos*. Cancer Research. 2008;**68**(20):8573-8581

[103] Chandrasekara A, Daugelaite J, Shahidi F. DNA scission and LDL cholesterol oxidation inhibition and antioxidant activities of Bael (*Aegle marmelos*) flower extracts. Journal of Traditional and Complementary Medicine. 2018;**8**(3):428-435

[104] Husain Khan T, Sultana S. Effect of *Aegle marmelos* on DEN initiated and 2-AAF promoted hepatocarcinogenesis: A chemopreventive study. Toxicology Mechanisms and Methods. 2011;**21**(6): 453-462

[105] Jatav S, Dwivedi P, Singh M, Sehra N, Mishra BB. Properties and Important Molecules of Medicinal Interest in Wood Apple (*Aegle marmelos*). In Synthesis of Medicinal Agents from Plants USA. Amsterdam, Netherlands: Elsevier; 2018. pp. 127-150

[106] Geetha T, Varalakshmi P. Anti-inflammatory activity of lupeol and lupeol linoleate in rats. Journal of Ethnopharmacology. 2001;**76**:77-80

[107] Patel PK, Sahu J, Sahu L, Prajapati NK, Dubey BK. *Aegle marmelos*: A review on its medicinal properties. International Journal of Pharmaceutical and Phytopharmacological Research. 2012;**1**(5):332-341

[108] Nigam V, Nambiar VS. Therapeutic potential of *Aegle marmelos* (L.) Correa leaves as an antioxidant and anti-diabetic agent: A review. International Journal of Pharma Sciences and Research. 2015;**6**(3):611-621

[109] Arul V, Kumaraguru S, Dhananjayan R. Effects of aegeline and lupeol-the two cardioactive principles isolated from the leaves of *Aegle marmelos* Corr. Journal of Pharmacy and Pharmacology. 1999;**51**:252

[110] Agrawal SS, Kumar A, Gullaiya S, Dubey V, Nagar A, Tiwari P, et al. Antifertility activity of methanolic bark extract of Aegle marmelos (L.) in male wistar rats. Daru. 2012;**20**(1):94

[111] Chauhan A, Agarwal M. Reversible changes in the antifertility induced by *Aegle marmelos* in male albino rats. Systems Biology in Reproductive Medicine. 2008;**54**(6):240-246

[112] Verma RS, Padalia RC, Chauhan A. Essential oil composition of *Aegle marmelos* (L.) Correa: Chemotypic and seasonal variations. Journal of the

Science of Food and Agriculture. 2014;**94**(9):1904-1913

[113] Baliga MS, Bhat HP, Pereira MM, Mathias N, Venkatesh P. Radioprotective effects of *Aegle marmelos* Corr: A concise review. The Journal Alternative and Complementary Medicine. 2010;**16**(10): 1109-1116

[114] Sharma PC, Bhatia V, Bansal N, Sharma A. A review on Bael. Natural Product Radiance. 2007;**6**(2):171-178

[115] Jagetia GC. Venkatesh P, Baliga MS. *Aegle marmelos* L. Corr inhibits the proliferation of transplanted Ehrlich ascites carcinoma in mice. Biology and Pharmaceutical Bulletin. 2005;**28**(1):58-64

[116] Jagetia GC, Venkatesh P, Baliga MS. Fruit extract of *Aegle marmelos* Corr protects mice against radiation-induced lethality. Integrative Cancer Therapies. 2004;**3**(4):323-325

[117] Sharma GN, Dubey SK. Ulcer healing potential of *Aegle marmelos* Corr fruit seed. Asian Journal of Pharmacy and Life Science. 2011;**1**(2):172-178

[118] Maity P, Hansda D, Bandyopadhyay U, Mishra DK. Biological activities of crude extracts and chemical constituents of *Aegle marmelos* L. Corr. Indian Journal of Experimental Biology. 2009;**47**(11): 849-861

[119] Simmons RA. Developmental origins of diabetes: The role of oxidative stress. Free Radical Biology and Medicine. 2006;**40**(6):917-922

[120] Upadhya S, Shanbhag KK, Suneetha G, Naidu BM, Upadhya S. A study of hypoglycemic and antioxidant activity of *Aegle marmelos* in alloxan induced diabetic rats. Indian Journal Physiology and Pharmacology. 2004;**48**(4):476-480

[121] Kamalakkannan N, Prince PSM. Hypoglycaemic effect of water extract of *Aegle marmelos* Corr: Fruit in streptozotocin diabetic rats. Journal of Ethnopharmacology. 2003;**87**(2/3): 207-210

[122] Birudu R, Pamulapati P, Manoharan S. Evaluation of biochemical changes in diabetic rats treated with *Aegle marmelos* (L.) methanolic leaf extract. Pharmacognosy Research. 2020;**12**:127

[123] Gohil T, Pathak N, Jivani N, Devmurari V, Patel J. Treatment with extracts of Eugenia jambolana seed and *Aegle marmelos* leaf extracts prevents hyperglycemia and hyperlipidemia in alloxan induced diabetic rats. African Journal of Pharmacy and Pharmacology. 2010;**4**:270-275

[124] Kamalakkannan N, Prince PSM. Antihyperlipidaemic effect of *Aegle marmelos* Corr fruit extract in streptozotocin induced diabetes in rats. Journal of the Science of Food and Agriculture. 2005;**85**:569

[125] Ilango K, Chitra V. Wound healing and anti-oxidant activities of the fruit pulp of *Limonia acidissima* Linn (Rutaceae) in rats, Trop. Journal of Pharmaceutical Research. 2010;**9**:223-230

[126] Pattanayak SP, Sunita P. Wound healing, anti-microbial and antioxidant potential of *Dendrophthoe falcata* (L.f) Ettingsh. Journal of Ethnopharmacology. 2008;**120**:241-247

[127] Veerappan A, Miyazaki S, Kadarkaraisamy M, Ranganathan D. Acute and subacute toxicity studies of *Aegle marmelos* Corr., an Indian medicinal plant. Phytomedicine. 2007;**14**:209-215

[128] Patel P, Asdaq SMD. Immunomodulatory activity of methanolic fruit extract of *Aegle marmelos* in experimental animals.

Saudi Pharmaceutical Journal. 2010;**18**:161-165

[129] Duraipandiyan V, Ayyanar M, Ignacimuthu S. Antimicrobial activity of some ethnomedicinal plants used by Paliyar tribe from Tamil Nadu, India. BMC Complementary and Alternative Medicine. 2006;**6**:35

[130] Chatterjee P, Chakraborty B, Nandy S. Review on Nephroprotective activity study by different plant extract. Advanced Journal of Pharmacie and Life Science Research. 2014;**2**(2):24-40

[131] Ahmad R, Mujeeb M, Ahmad A, Anwar F. Ameliorative effect of *Aegle marmelos* leaves extract against cisplatin-induced nephrotoxicity and oxidative stress. Bangladesh Journal of Pharmacology. 2016;**11**:101-109

[132] Arul V, Miyazaki S, Dhananjayan R. Mechanisms of the contractile effect of the alcoholic extract of *Aegle marmelos* Corr. On isolated Guinea pig ileum and tracheal chain. Phytomedicine. 2004;**11**:679-683

[133] Laphookhieo S, Phungpanya C, Tantapakul C, Techa S, Tha-In S, Narmdorkmai W. Chemical constituents from *Aegle marmelos*. Journal of the Brazilian Chemical Society. 2011;**22**: 176-178

[134] Sankari M, Chitra V, Silambujanaki P, Raju D. Anticonvulsant activity of ethanolic extract of *Aegle marmelos* (leaves) in mice. International Journal of PharmTech Research. 2010;**2**:640-643

[135] Behera J, Mohanty B, Ramani Y, Rath B, Pradhan S. Effect of aqueous extract of *Aegle marmelos* unripe fruit on inflammatory bowel disease. Indian Journal of Pharmacology. 2012;**44**: 614-618